A History of Disease in Ancient Times

Philip Norrie

A History of Disease in Ancient Times

More Lethal than War

palgrave
macmillan

Philip Norrie MBBS, MA, MSc, MSocSc (Hons), PhD, MD
Conjoint Senior Lecturer, Faculty of Medicine,
University of New South Wales
Sydney, Australia

Faculty Affiliate, School of Public and Community
Health Sciences, College of Health Professions
and Biomedical Sciences, University of Montana,
Missoula, MT, USA

ISBN 978-3-319-28936-6 ISBN 978-3-319-28937-3 (eBook)
DOI 10.1007/978-3-319-28937-3

Library of Congress Control Number: 2016939237

Printed on acid-free paper

This Palgrave Macmillan imprint is published by Springer Nature
The registered company is Springer International Publishing AG Switzerland

This book is dedicated to the people, forgotten by history, who died from the epidemics that helped end the Bronze Age in the Near East.

Foreword

Dr. Norrie's excellent work–*A History of Disease in Ancient Times*–is one of the most important and novel contributions to the history and theory of epidemiology in decades. The book emphasizes a crucial concept that historians seem to have missed. This is that infectious epidemic diseases would have been a crucial explanatory variable for the cyclic changes of the Bronze Age, including the collapse of the major civilization of that era.

Dr. Norrie's work captures evidence for epidemic impacts from a variety of sources–historic, archeologic, linguistic, medical, social, anthropological and economic–and makes a convincing argument for including epidemiology in discussions of Bronze Age changes. Such changes have been poorly understood and controversial before this contribution. What Norrie has done that is especially significant is to introduce the epidemiogic notion of the "web of causation" to patterns of change in the civilizations of that era. When epidemiology is added to the mix of explanatory theories, the picture of Bronze Age change and collapse becomes more focused and reasonable. This is a major intellectual accomplishment. Dr. Norrie provides multiple examples of this interaction, some well-known and some less well-known, up to the struggles of ancient Rome with its arch-enemy Carthage.

As Dr. Norrie notes, "Disease has killed more people, over the ages, than all the wars, famines and other disasters put together." This is a well-known axiom in modern public health and epidemiology, and is a concept commonly taught in modern public health curricula in epidemiology, relying on historic examples and data, which clearly illustrate this point and are legion. What Norrie then asks is a deceptively simple and immensely powerful question—Why should this have been different in Antiquity?

The answer is that it would not have been different. It would behoove historians and other aficionados of "the Antique" to consider their theories of evolution, change and decline within the possibilities of a web that metaphorically captures system change, and allows biological events at the population level to have meaning.

The great Charles Rosenberg, in his classic *The Cholera Years*, notes one of the central axioms of epidemiology: "A disease is no absolute physical entity but a complex physical construct, an amalgam of biological state and social definition." In recent years the public health model has come to a great appreciation of this concept. Philip Norrie deftly incorporates this concept throughout his book. It is a concept that, beyond modern utilization in the age of the Ebola outbreak in West Africa, also would have had meaning in the Bronze Age epidemics, where severity and degrees of infection would have been variable and susceptible to interpretation.

This book will provide excellent service to those who practice medicine, anthropology, epidemiology and ancient history. It provides a wonderful summation of historical material, an interesting and highly plausible new theory, and would be an excellent (and needed) supplemental reading for other new works, such as Eric Cline's "1177 B.C.—The Year Civilization Collapsed". I very much appreciate Dr. Norrie's efforts to re-think the Bronze Age with an open and creative mind.

Craig A. Molgaard BA, MDH, PhD
Professror and Chair
School of Public and Community
Health Sciences
College of Health Professions
and Biomedical Sciences
University of Montana
Missoula, MT, USA

Acknowledgements

I would like to thank the following people for helping me research and write this book.

Professor Mark Ferson from the Medical Faculty at the University of New South Wales, my Doctor of Medicine thesis Supervisor and Dr. Susan Hardy from the History Department at the University of New South Wales, my Doctor of Medicine thesis Co-Supervisor for their guidance and support throughout the whole thesis process.

Dr. Julian Henwood PhD, a good friend, for his editorial skill morphing my Doctor of Medicine thesis into its final form.

Finally Mrs Michelle Catterick, also a good friend, for all her invaluable clerical and computer skills; which were much needed during the long process of thesis and book production.

Dr. Philip Norrie, Sydney

Contents

List of Figures

List of Tables

CHAPTER 1

Introduction

"Historians find war exciting and pestilence dull; they exaggerate the effects of the former and play down the latter". (Grove and Rackham 2001)

This book will explore the way in which infectious diseases affected the course of ancient Mediterranean and Near Eastern history. The general hypothesis is that most major changes in the Ancient World were precipitated by infectious disease epidemics, for example, the end of the Mycenaean Palatial Empires, the end of the Hittite Empire and the end of the Near Eastern Bronze Age in c.1200 BCE; Egypt's withdrawal from the Levant, the fall of Athens, the end of Carthage, the fall of the Western Roman Empire, the invasion of Islamic forces into Europe during the Plague of Justinian; and even the end of the Sumerian and Indus Valley Civilizations. And if the Ebola virus gets out of control—it may be 1200 BCE all over again!

In ancient times it would have been a daily struggle for survival either through avoiding or fighting against infectious diseases: sepsis from *Staphylococcus aureus* or tetanus from a graze; childhood infectious diseases such as whooping cough, diphtheria and measles; the infections from contaminated food and water; let alone major infections such as leprosy, tuberculosis and malaria or the big two killers—smallpox and bubonic plague.

The plea of this book is that, in future, ancient historians consider the potential role of infectious disease in the histories they research and

P. Norrie, *A History of Disease in Ancient Times*,
DOI 10.1007/978-3-319-28937-3_1

subsequently write. If disease is not considered and is ignored, as is the current situation, then the resultant history may be incomplete and thus flawed; because you cannot administer or feed let alone defend your empire if your citizens are dying en masse due to an infectious disease epidemic.

The latest book about the end of the Bronze Age, thus covering a similar area in ancient history as this book, is Professor Eric Cline's book *1177 B.C.–The Year Civilization Collapsed.*[1] It gives a very good summary of the current state of conventional historical thinking about the end of the Bronze Age in the Near East; but it fails to discuss the role of infectious disease. This book will hopefully fill that void. Professor Cline and I have had profitable discussions about this topic via email. He agrees with me that it would be useful to consider this and suggested that I collect more evidence and proof.

Some ancient historians and archaeologists might argue that what I have to say is speculative, with an added "what if" factor. However, they cannot ignore that what I am providing is a plausible alternative hypothesis—that infectious disease epidemics could have ended the Near Eastern Bronze Age, along with the other empires mentioned previously. This attitude is understandable and is not unusual when new ideas challenge established ideas. To use an appropriate analogy from medical history—when Ignaz Semmelweis said that puerperal fever was caused by the infected hands of the doctors, the obstetric establishment scorned and criticized him, rejecting his ideas, until he was proven correct.

Throughout history disease has killed many more people than have been killed in wars.

> More lethal than any firearms, viruses and bacteria have accompanied soldiers and traders around the world, and the epidemics they caused have changed the course of history.[2]

Using mainly the vast quantity of secondary literature on this topic, this book, then, will show that a historiographical approach to looking at the impact of disease will not only produce a much needed summary and synthesis of the current literature but, in doing so, will also present a coherent description and explanation of the influence of disease on civilization.

This book will show how a study of the medical history of particular places and periods can expand the understanding of historical events.

"Historians study change"[3] and the key to this change is causation, "how chains of interaction become pathways of historical development",[4] this is, what factors were involved to make things turn out the way they did. Diseases, as will be demonstrated, are links in these chains.

This book will also explore the issue of disease causation, examining both the understanding of disease processes in the past and also what our current etiological approach might be, while maintaining an awareness of the dangers of retrospective diagnosis and the diagnostic limitations due to poor records and the depth of medical knowledge in different periods of history. The ancient texts talk of "plagues" meaning, in current western terminology, an infectious disease affecting and killing large numbers of people at the same time; currently "plague" is understood to refer to many different specific diseases such as smallpox or Ebola virus or the classic Bubonic plague but in ancient times it was used as a general descriptor. It is not possible to be sure which specific organism was responsible for "plague" in early times. It is also important to remember that infectious diseases evolve over the centuries while populations develop immunity; so what would have caused devastation in ancient times in a virgin population due to its virulence and the lack of resistance in the population (such as the measles virus) may be a lot milder today with populations with improved immunity.

The causes of historical events are multifactorial but medical aspects must always be considered as a major contributing factor. As Prof. Roy Porter has noted:

> Writing this book … has bought home the collective and largely irremediable ignorance of historians about the medical history of mankind. Perhaps the most celebrated physician ever is Hippocrates yet we know literally nothing about him. Neither do we know anything concrete about most of the medical encounters there have ever been. The historical record is like the night sky: we see a few stars and group them into mystic constellations. But what is chiefly visible is the darkness.[5]

This book will try to fill in some of the gaps and thus reduce the darkness. The findings of medical history have shown how much disease and resulting death have greatly influenced events in the past, most basically because it is disease that determines the fitness of a society and whether its citizens, both great and small, who make up that society and its history, live or die and how they lived or died. Disease has killed more people, over the ages, than all the wars, famines and other disasters put together. For example, smallpox alone killed an estimated 300 million people in the twentieth century before it was eradicated. This is approximately three times more than all the people killed in all the wars in the twentieth century.[6]

Traditional history tended to concentrate on the politics, military campaigns, royalty, Empires, religions, science, economics, art and exploration, but not medicine. Recently, however, this situation has changed. The change was led by such prolific writers as the late Prof. Roy Porter of University College in London, Prof. Charles Rosenberg from Harvard University in Boston and Emeritus Prof. Barry Smith of the Australian National University in Canberra. Nevertheless, the incorporation of medical issues into mainstream history still lags behind incorporation of the issues mentioned above. For example, the history section in the *Encyclopedia Britannica* states, in the Macropedia section "Special aspects of history are also dealt with in articles under the classifications of military affairs, economics, law, literature, sciences, art, philosophy, religion and political science".[7] Note medical history is not mentioned, while one of Australia's most respected historians, Prof. Geoffrey Blainey, shows a similar lack of recognition of medical history's importance when in the Preface to his book *A Short History of the World* (669 pages long) he talks about putting emphasis on the important things that "have done so much to shape the world"[8] namely: technology and skills, religion, geographical factors or "the tyranny of distance", eating habits, work and empires. There is no mention of medicine or the effect of disease on events in the past although he does note, albeit in a strangely dismissive tone that "such profound and symbolic events as the invention of anaesthetics, surely call for at least one paragraph; but they receive scant or no attention in this book".[9] However as the late Prof. Roy Porter, an equally eminent historian noted "Medicine has played a major and growing role in human societies and for that reason its history needs to be explained so that its place and powers can be understood".[10]

Prof. Roy Porter (1946–2002) was a British medical historian who eventually became Professor of Social History and the Director of the Wellcome Institute for the History of Medicine at University College, London. He wrote eighteen books on the history of medicine.

Medicine contributes much to social progress because society cannot advance and progress as rapidly if the people in it are unhealthy. Samuel Johnson, a notable eighteenth century "social commentator", once referred to the medical profession as "the greatest benefit to mankind"[11] and it is difficult to see how a complete history of any civilization could be contemplated without its inclusion.

This book will offer new perspectives, possibilities, points of view and insights into ancient Near Eastern history: first, by updating current knowledge about the end of the Bronze Age and the Hittite Empire for example; second, by expanding the infrequently recognized effects of disease in past periods, that is, debilitation and subsequent death in the Hittite Army; third, by showing how disease influenced major historical events such as the end of Carthage. This book will also construct a synthesis of existing knowledge concerning reasons for the ending of the Near Eastern Bronze Age. All of these examples will be expanded upon later in this book.

Finally, this book provides an insight into the role of disease in the history of the ancient Mediterranean and Near East showing its significance as a major contributing causal factor in the end of the Near Eastern Bronze Age, the end of the Hittite Empire and the end of Carthage and why it should be given more recognition and focus in the future by historians.

Notes

1. Cline, E. H., *1177 B.C. – the year civilization collapsed*, Princeton and Oxford: Princeton University Press, 2014).
2. Daniels, P.S. and Hyslop, S.G., (2003) *National Geographic Almanac of World History*, (Washington D.C.: National Geographic Society, 2003), 24.
3. Hobsbawn, E., *On History*, (London: Weidenfield Nicolson, 1997), 43.
4. Jervis, R., *Counterfactuals, Causation and Complexity in Tetlock, P.E. and Belkin, A. (eds) Counterfactual Thought Experiments in World Politics: Logical, Methodological and Psychological Perspectives*, (New Jersey: Princeton University Press, Princeton, 1996), 310.
5. Porter, R., op. cit. (1999), Introduction, 13.
6. Oldstone, M.B.A., (2000) *Viruses, Plagues and History*, (Oxford: Oxford University Press, 2000), 27.
7. Goetz, P.W., op. cit., Vol.5, 949.
8. Blainey, G., *A Short History of the World*, (Victoria: Penguin Group Australia, Camberwell, 2001), Preface.
9. Ibid.
10. Porter, R., *The Greatest Benefit to Mankind*, (New York: W.W. Norton and Co., 1999) Introduction.
11. Ibid, Introduction, 4.

CHAPTER 2

Common Diseases in Ancient Times

Introductory Background History

Europe and its environs, in the times before the Common Era [BCE], was focused around the Mediterranean Sea, North Africa, Egypt and the Near and Middle East, where civilization began in the Fertile Crescent (see Image 2.1).

The first civilization (an advanced stage or system of human social development)[1] was in Mesopotamia (Iraq) along the Tigris and Euphrates rivers, dating from about 3500 BCE.[2] This civilization reached its zenith as the Akkadian or Agade Empire from 2360 to 2230 BCE[3] or from 2371 to 2230 BCE,[4] especially under Sargon the Great around 2279 BCE.[5] This was followed by the Egyptian civilization along the Nile River, which first appeared about 3100 BCE[6] and reached its zenith as the Middle Kingdom of Egypt from 1980–1630 BCE.[7] During the eighteenth Dynasty of the New Kingdom the Israelites left Egypt (the Exodus) to establish their own state in the Near East in Palestine. Disease played a role in the Exodus and in the subsequent early history of Israel when it was trying to establish itself, especially in its war with the Philistines. Egypt's glory ended with the end of the 33rd Dynasty in 30 BCE.[8] The next European civilization was the Minoan Civilization in Crete starting from about 2100 BCE.[9]

After the collapse of the Akkadian Empire there followed, in Mesopotamia, a period of three centuries of chaos that created a power vacuum. Eventually two new empires emerged —the Assyrian Empire in

P. Norrie, *A History of Disease in Ancient Times*,
DOI 10.1007/978-3-319-28937-3_2

Northern Mesopotamia, based around the cities of Ashur and Nineveh and the Babylonian Empire in Southern Mesopotamia based around its capital city Babylon. Two centuries later the Hittites emerged from Anatolia (Turkey) in about 1700 BCE and their empire lasted until about 1200 BCE.[10] Disease played a significant role in the demise of the Hittite Empire. Disease also helped weaken the Assyrian Empire, which ended in 612 BCE when its capital Nineveh was invaded by the Chaldeans[11] and similarly the Babylonian Empire ended in 539 BCE when its capital Babylon was captured by the first of the new Persian (Iran) Empire kings, Cyrus the Great.[12] The history of Europe, for the remainder of the pre-common era period, is dominated initially by the rivalry of the Greek and Persian Empires and then by the rivalry of the Empires of Carthage and Rome.

The Persian Empire started with Cyrus the Great in 560 BCE and ended with the death of Darius III in 330 BCE[13] after he had been defeated by Alexander the Great of Macedonia and Greece. During that time it competed with and eventually invaded Greece under Xerxes I, with the largest army ever assembled in ancient times, but disease, in the form of dysentery, helped

Image 2.1 Map of Ancient Mediterranean Sea, Near East and Middle East (© Dorling Kindersley/Thinkstock)

destroy this army. Disease, in the form of malaria, also helped Alexander the Great come to power by conquering Greece and disease ended his Empire when he, most probably, died of West Nile viral encephalitis. Disease also was the main factor in the fall of Athens, with the Plague of Athens, during the Peloponnesian War.

Carthage, on the North African coast, was founded in the ninth century BCE by the Phoenicians of Tyre. During Roman times the eastern spread of the Carthaginian Empire was stopped in Sicily by diseases such as measles, influenza and typhus several times—the last being during the Second Punic War. Disease also helped the Carthaginian General Hannibal defeat Roman forces and occupy Northern Italy thus threatening Rome's hold on Italy itself.

The Roman Empire was subject to recurrent infectious disease epidemics, such as smallpox, leprosy and malaria, throughout its history and ended in 476 in the grips of a falciparum malaria epidemic.

The author of this book has a great interest in medical history and a long held belief that the role of disease in history has for too long been neglected by historians. Furthermore, any historical analysis of any subject is incomplete, hence inadequate, if the possible ramifications of the effect of disease on that topic have not been considered.

The end of the Bronze Age and the demise of the Hittite Empire is a good example of this historical neglect. The end of the Bronze Age was called "the catastrophe" and was characterized by its short time frame, mass migrations of people (including the Sea People) and abandonment of cities such as, the Hittite Empire's capital Hattusa. Could the short time frame of just 50 years from c.1200 BCE to c.1150 BCE, be due to an infectious epidemic or epidemics, could the mass migration of populations be due to people fleeing disease, could the abandonment of whole cities be due to people fleeing disease, and could the invasion by the "Sea People" raiders be because they too were fleeing disease in their homelands and were looking for a new place to live? This book proposes that disease in the form of an infectious epidemic or epidemics could be a possible major cofactor in the end of the Bronze Age. The only time that disease has ever been mentioned as a possible cause for the end of the Bronze Age was by Tom Slattery, in his book *The Tragic End of the Bronze Age—A Virus Makes History*, where he only talks about smallpox; Eric Watson-Williams and Lars Walloe in their two articles also mention bubonic plague. This book will additionally propose other diseases such as tularemia, malaria and dysentery, to name but a few.

Disease "fits" the scenario of the end of the Bronze Age so well: the end was rapid; it involved a large area where something had to be free to cross borders; it was also characterized by mass migrations of populations whether they be the innocent general population or raiders/invaders such as the "Sea Peoples"; and finally any explanation for the end of the Bronze Age has to account for and explain the abandonment of whole cities. If a city was to be raided or invaded it would show signs of destruction and subsequent occupation. However a lot of cities, from this period don't show any of this, they show only abandonment. A disease or a series of different diseases in different places would be rapid, would cross borders, would cause people to flee in mass migrations and would also cause people to abandon their city, leaving it intact and unscathed. This book proposes a multicausal explanation for the end of the Bronze Age, which had disease as the major cause.

There is good evidence from the study of tree rings that the Near East suffered from drought at the time of the end of the Bronze Age. The cause of the drought (a natural occurrence, as a result of volcanic eruptions, as a result of comet activity or a combination of all three) is unclear. Earthquake activity and resultant tsunamis would have only added to the misery. The drought led to famine that in turn made people more prone to disease. Disease in the form of a devastating infectious plague or plagues would have killed so many people that all the empires would have been unable to feed, trade, administer or govern themselves, let alone defend the empire from invasion by neighboring states, such as the Kaska people in the case of the Hittite capital of Hattusa, or from invasion by "Sea Peoples".

The populations in the major cities had three options at the end of the Bronze Age: (1) flee by abandoning their city and then become part of the mass migration seen at this time; (2) burn their city to destroy the disease and again become part of the mass migration; or (3) stay and fight the invaders and risk losing the battle and seeing their city destroyed.

Changes in warfare and royal family infighting could have been contributing factors. The disruption of trade routes by the Sea Peoples affected trade, especially in wine and food, and along with economic factors and general systems collapse are seen as a "consequence of" and not a "cause of" the end of the Bronze Age due to disease.

The Diseases that Could Have Affected Ancient Times

The diseases that could have caused change and devastation in ancient times include:

Zoonoses

A zoonosis is any infectious disease that can be transmitted from animals (both wild and domestic) to humans.[14] The Near East in 1200 BCE was dominated by rural agricultural activity where peasants lived with wild and domesticated animals, hence having contact with them all the time.

Animals such as bats, cats, cattle, dogs, fish, fleas, flies, geese, goats, horses, lice, mice, mosquitoes, pigs, rabbits, rats, rodents, sheep, snails, ticks and wolves, were present in the Near East c.1200 BCE. These animals caused many different zoonoses in the Near East such as, anthrax, brucellosis, bubonic plague (see Image 2.2), cholera, encephalitis, *Escherichia coli (E. coli)*, rabies, salmonellosis, tularemia, typhus and West Nile virus. Malaria is not a zoonosis, even though it is transmitted by a mosquito, because it depends on the human to be a host for part of its lifecycle.

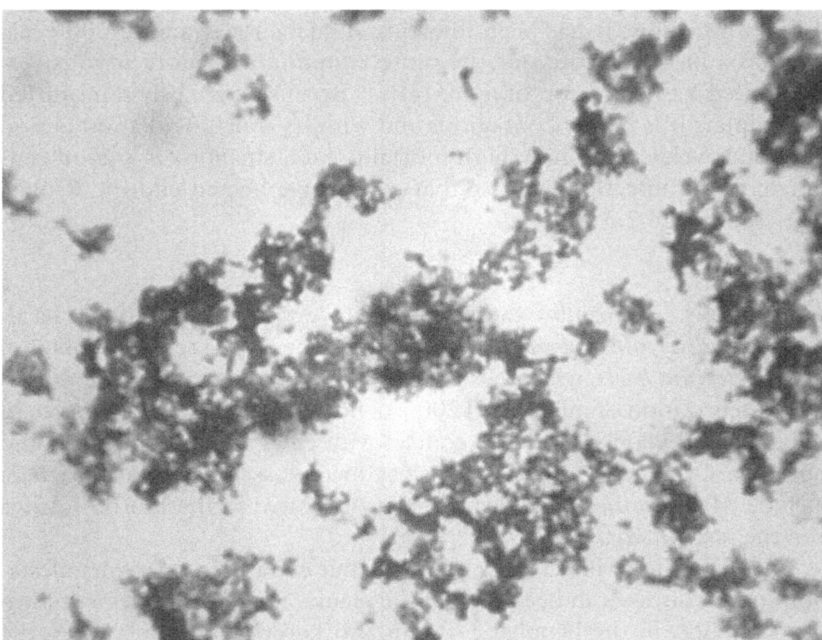

Image 2.2 Photograph of *Yersinia pestis* (bubonic plague) (© Duncan Smith/Photodisc/Thinkstock)

Of these zoonoses, brucellosis, *E. coli*, West Nile virus and rabies would not be generalized enough, nor sustain an epidemic long enough to cause the end of the Bronze Age. The other zoonoses such as bubonic plague, cholera, salmonellosis, typhus, encephalitis and tularemia would be capable of being spread over large areas and large populations, as well as lasting long enough to cause significant harm.

Leprosy or Hansen's Disease

Leprosy is a chronic disease caused by the bacteria *Mycobacterium leprae* and has been infecting man for over 4000 years.[15] As a chronic slowly evolving disease it would have been too slow to develop to be the devastating epidemic that contributed strongly to the ending the Hittite Empire or the Bronze Age.

Smallpox

Smallpox is caused by a virus, which is transmitted human to human via droplet infection. It has been infecting humans for some 12,000 years and has a mortality rate, for the more common and more serious version called Variola major, of up to 60%.[16] Because it is easily transmitted via droplets it is highly contagious and with its short incubation period of about twelve days and a high mortality rate, smallpox is a good candidate for an infectious disease that could have helped end the Bronze Age (see Image 2.3).

Tuberculosis

Tuberculosis is mainly caused by the aerobic non-motile bacillus *Mycobacterium tuberculosis* and sometimes by the less common *Mycobacterium bovis*, which has been eliminated as a major cause due to the pasteurization of milk.[17] In 1200 BCE however it was a major source of the disease as people had more contact with cows and pasteurization did not exist. Other less common forms of tuberculosis causing mycobacteria include *M. africanum* in Africa, *M. canetti* located in the Horn of Africa and the rare *M. microti*.

The main site of infection is the lungs but extra-pulmonary manifestations of the disease can been seen in the pleura, in the meninges as a form of meningitis, in the lymphatic system also known as scrofula when the cervical lymph nodes are involved, in the urinary system and in bones especially in the spine when it is known as Pott's Disease (named after the

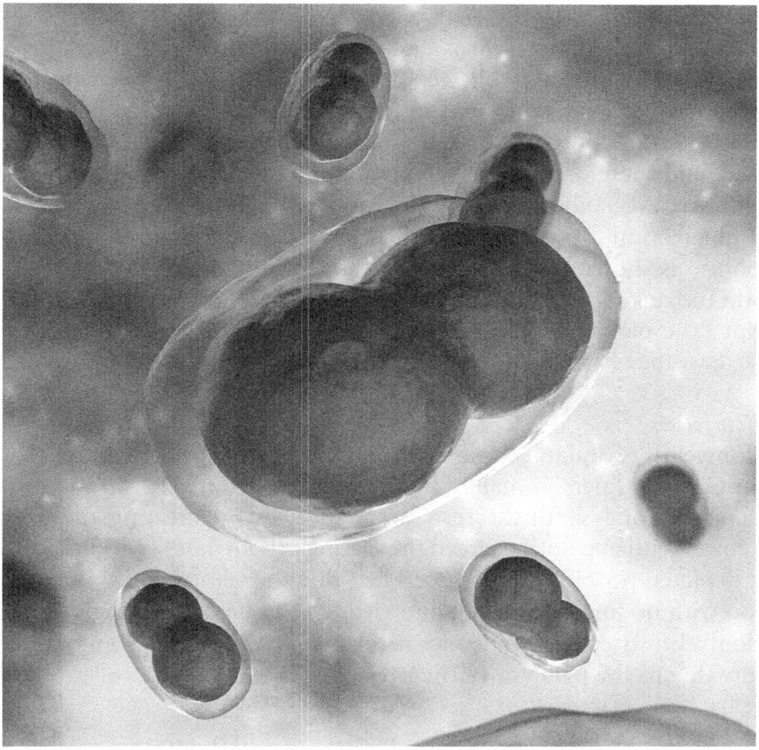

Image 2.3 Photograph of *Variola* (smallpox virus) (© decade3d/iStock/Thinkstock)

St. Bartholomew's Hospital surgeon Percival Pott (1714–1788) who first described it).

Tuberculosis is highly contagious and is transmitted by droplet infection when an infected person coughs, sneezes, spits, sings or speaks. It has been found in the remains of bison from 17,000 years ago[18] and in human skeletal remains from 6000 years ago. Even Egyptian mummies from 3000 to 2400 BCE have shown signs of Pott's disease.[19] The Greek physician Hippocrates in c.460 BCE described it as the most widespread disease that was nearly always fatal; hence it would be a good candidate for the disease that was a contributory factor in ending of the Bronze Age and the Hittite Empire.

Influenza

Influenza is caused by a virus and was described by Hippocrates 2400 years ago. The "Spanish Flu" epidemic of 1918–1919 killed up to 100 million people worldwide making it the single most lethal pandemic in history;[20] so it's potential as a significant actor should not be underestimated, and therefore a good candidate for a disease influential in the end the Bronze Age and the Hittite Empire.

Influenza, in a population with no previous exposure could be catastrophic, as shown by first contact death rates of up to 90% in the New World Indians and the Pacific Islanders. After the Amarna period of Ancient Egypt "the pandemic that followed this period throughout the Ancient Near East may have been the earliest recorded outbreak of influenza".[21]

Poliomyelitis

Poliomyelitis is viral infectious disease caused by the polio virus and spread from person to person mainly by the oral-fecal route. It can cause death by causing paralysis of the respiratory muscles by affecting the nervous system. Egyptian paintings from the eighteenth Dynasty (1403–1365 BCE) had depicted people with withered legs due to poliomyelitis.[22] In a population with no immunity, the polio virus would be devastatingly lethal. As poliomyelitis occurs in epidemics many people would die from respiratory failure due to the paralysis of their respiratory muscles.

Measles

Measles is a viral infection of the upper and lower respiratory system[23] caused by the *morbillivirus*. Like smallpox and influenza, measles in an unprotected population is deadly as shown by its high mortality rates in the New World and the Pacific when they were first occupied by Europeans.

Malaria

Malaria is an infectious disease caused by plasmodium parasites that enter the body's blood stream from a bite by the Anopheles mosquito.[24] Malaria has been infecting humans for over 50,000 years. There are several malaria-causing plasmodia namely *P. falciparum*, *P. malariae*, *P. ovale*, *P. vivax* and *P. knowlesii*. *P. vivax* causes the most infections but *P. falciparum* is the most deadly causing up to 90% of all deaths. Malaria would have been present in Egypt and the southern Levant in 1200 BCE.

Typhoid Fever

Typhoid fever is caused by ingesting food or drinks contaminated by the bacterium *Salmonella typhi*.[25] With the poor sanitation conditions present at the end of the Bronze Age it would have been a common infection causing severe high fevers, intestinal hemorrhage, encephalitis, delirium, abscesses and finally death. Typhoid fever is thought to have been the cause of the Plague of Athens (c. 430–424 BCE), which killed one-third of the population of Athens.

Dysentery

Dysentery is an infection of the colon that causes severe diarrhea, bleeding from the bowel, high fever and eventually fatal dehydration. There are two types due to the different infective agents. The first type is amoebic dysentery caused by the amoeba *Entamoeba histolytica* and the second type is bacillary dysentery caused by the bacterium shigella, which results in severe dysentery.[26] The poor sanitation conditions in the Near East c.1200 BCE would have helped spread all these diseases.

Now we shall examine in the next chapters how these infectious diseases may have affected ancient history on the Mediterrean sea, Near East and Middle East, and thus change history!

Notes

1. Pearsall, I. (Ed)., *Concise Oxford English Dictionary, Tenth Edition, Revised* (Oxford University Press, 2002) 261.
2. Roberts, J.M., *Ancient History from the First Civilisations to the Renaissance* (London: Duncan Baird Publishers, 2004) 69.
3. Danzer, G.A., *Atlas of World History* (Kent, U.K: Grange Books, Rochester, 2002) 23.
4. Saggs, H.W.F., *The Babylonians* (London: The Folio Society, 2005) 415.
5. Farrington, K., *Historical Atlas of Empires from 4000 BC to the 21st Century*, (London: Mercury Books, 2003) 11.
6. Roberts, op.cit., 69.
7. Ibid, 107.
8. Ibid.
9. Ibid, 69.
10. Gurney, O.R., *The Hittites*, (London: The Folio Society, 2005) 205.
11. Saggs, op.cit., 419.
12. Ibid.
13. Cook, J.M., *The Persians* (London: The Folio Society, 2005) 350.

14. Zoonoses, http://www.dpi.nsw.gov.au/biosecurity/animal/humans/zoonoses. Accessed 7 December 13.
15. Leprosy, http://www.betterhealth.vic.gov.au/bhcv2/bharticles.nsf/pages/Leprosy_explained. Accessed 7 December 13.
16. Smallpox, http://www.infoplease.com/cig/dangerous-diseases-epidemics/smallpox-12000-years-terror.html. Accessed 10 October 11.
17. Tuberculosis, http://www.essortment.com/diseases-illnesses-history-tuberculosis-21066.html. Accessed 3 October 11.
18. Rothschild, B.M., Martin, L.F., Lev, G., et al., "*Mycobacterium, tuberculosis Complex DNA, from an extinct bison dated 17,000 years before the present*", (Clinical Infectious Diseases 33 (3), 2001), 305–11.
19. Zink, A., Sola, G., Reischl, U., Grabner, W., Rastogi, N., Wolf, H., Nerlich, A., "*Characterisation of Mycobarterium tuberculosis complex DNAs from Egyptian Mummies by Spoligotyping*", (Journal of Clinical Microbiology 41(1), 2003), 359–67.
20. Influenza, http://www.health.howstuffworks.com/diseases-conditions/cold-flu/history-of-influenza.htm. Accessed 7 December 13.
21. Ancient Egypt Outline Akhenaten on http://www.ancientegyptianline.co.uk/akhenaten.html. Accessed 3 December 2013.
22. Paul, J.R., *A History of Poliomyelitis*, (Yale Studies in the history of science and medicine, New Haven, Connecticut: Yale University Press, 1971) 16–18.
23. Measles, http://www.betterhealth.vic.gov.au/bhvc2/bharticles.nsf/pages/Measles. Accessed 7 December 13.
24. Malaria, http://www.betterhealth.vic.gov.au/bhvc2/bharticles.nsf/pages/Malaria. Accessed 7 December 13.
25. Typhoid, http://www.betterhealth.vic.gov.au/bhvc2/bhartivles.nsf/Typhoid_and_paratyphoid. Accessed 7 December 13.
26. Dysentery, http://www.who.int/topics/dysentery/en/. Accessed 7 December 13.

CHAPTER 3

How Disease Affected the History of the Egyptian Empire

Introduction

The Egyptian Empire was one of the great empires of the ancient world (see Image 3.1), if not the greatest. Egyptian history usually begins with the Early Dynastic Period (dynasties one and two) beginning c.3100 BCE (±150 years) with the reign of Menes, according to the classification by the Egyptian priest Manetho. Prior to this time were the Pre-Dynastic Kings, of which little is known. After the Early Dynastic Period came the Old Kingdom (dynasties three to seven) from c.2700 BCE, then the First Intermediate Period (dynasties eight to ten), followed by the Middle Kingdom (dynasties eleven and twelve) from c.2134 BCE and then the Second Intermediate Period (dynasties thirteen to seventeen). The New Kingdom (dynasties eighteen to twenty) began c.1575 BCE and was followed by the final period—the Late Dynastic Period (dynasties twenty one to thirty one) from c.1087 BCE until the end of the Egyptian Empire in 332 BCE when it was conquered by Alexander the Great. During this time it consolidated the Kingdoms of the Upper and Lower Nile and also expanded into the Levant (Palestine and Syria) (see Image 3.2).

The Egyptians developed a very complex medical system, which included a highly refined pharmacopeia and invasive surgical procedures such as trephining (drilling holes through the skull and into the brain).

P. Norrie, *A History of Disease in Ancient Times*,
DOI 10.1007/978-3-319-28937-3_3

Image 3.1 Photograph of Pyramids of Giza and Sphinx (© heckepics/iStock/Thinkstock)

There are seven Egyptian Medical Papyri in existence:
Kahun—a gynecological papyrus written c1900 BCE;
Edwin Smith—papyrus written c.1550 BCE, but the original author was an experienced physician from c.2200 BCE;
Ebers—papyrus written c.1500 BCE;
Hearst—papyrus written c.1500 BCE;
London—papyrus written c.1350–1100 BCE;
Berlin—papyrus written c.1350–1100 BCE; and
Brugsch—minor papyrus written c.1350–1100 BCE;

From these papyri an extensive knowledge of Egyptian medical practice can be gained. They document hundreds of specific prescriptions, including diverse combinations of medicines (polypharmacy). They contain evidence that they been copied from earlier papyri dating as far back as 3400 BCE. These papyri also show a consistency of prescribing: prescriptions that use plants, herbs, minerals and animal parts dissolved or mixed mainly in wine but also in beer, honey, water, milk and oil. These various solvents would have also made the medicines more palatable.

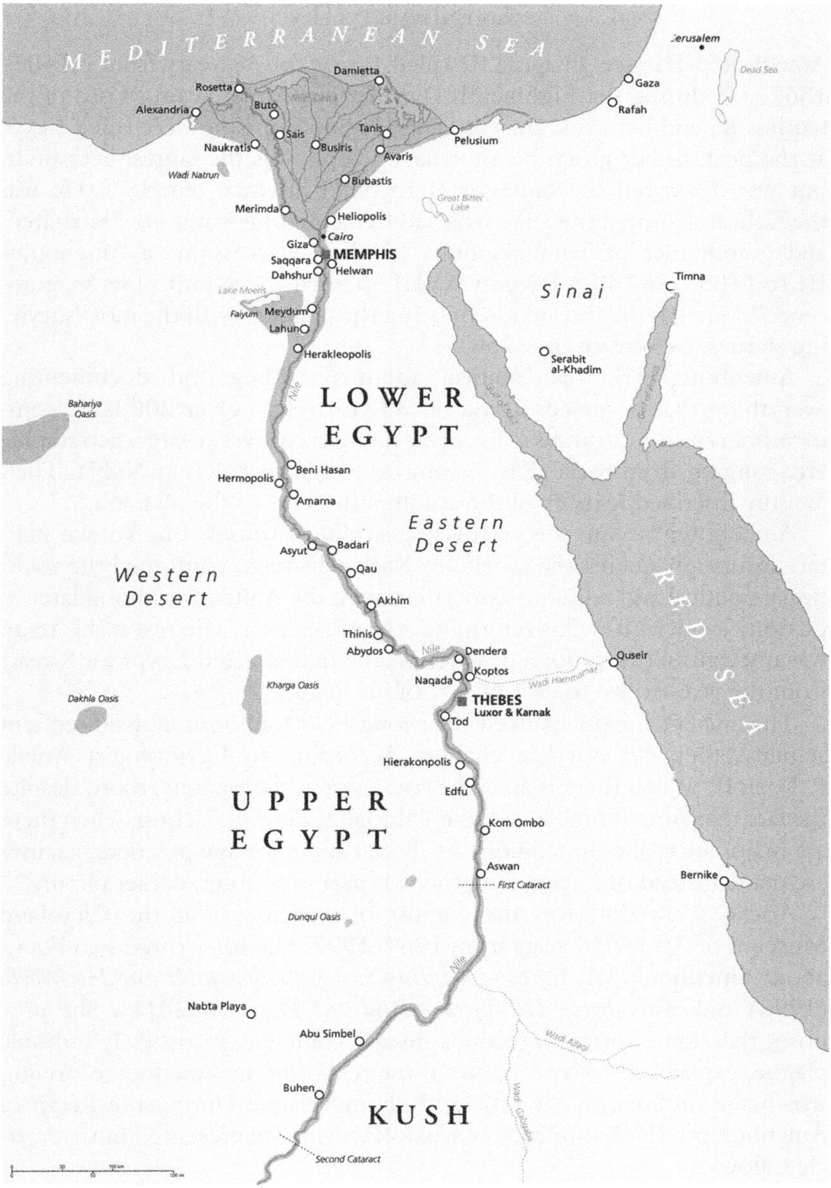

Image 3.2 Map of Ancient Egypt (© PeterHermesFurian/iStock/Thinkstock)

Amenhotep III

Amenhotep III (see Image 3.3) ruled Egypt for 38 years from c.1405–1367 BCE[1] during the Eighteenth Dynasty. His name meant "Lord of the truth is Ra and he was known as 'The Magnificent King'. He ruled Egypt at the peak of her glory, his mortuary temple was the largest ever built, but was destroyed by Ramesses II to build his own temple".[2] He was the father of Amenhotep IV, who later changed his name to Akhenaten, and grandfather of Tutankhamun. "With the accession of Amenophis III (c.1405–1367 BC) Dynasty XVIII attained the zenith of its magnificence".[3] He has the distinction of being the pharaoh with the most surviving statues: a count of over 250.

Amenhotep III was fanatical about recording and documenting everything that occurred during his 38-year reign. Over 200 large commemorative stone scarabs have been discovered over a large geographic area ranging from Syria (Ras-Shamra) through to Soleb in Nubia. Their lengthy inscribed texts extol the accomplishments of the pharaoh.[4]

Amenhotep's reign was relatively peaceful, with only one known military campaign against the rebellious Kush. His reign continued the traditional political and religious association with the Amun priesthood later to be done away with by his reforming son Akhenaten. The rest of his reign was uneventful except for a possible plague that affected Egypt for 8 years from the year twelve to year twenty of his reign.

The ancient Egyptians liked their long-held traditions and subsequent stability, they did not like change. According to Egyptologist Arielle P. Kozloff "When there is an eight year lapse in the written record despite the fact that Amenhotep loved to memorialize his every action, when there are major anomalies in religion, art, burial and marriage practices, excuses are made instead of recognizing them as part of a larger, darker picture".[5]

Arielle P. Kozloff was the curator of ancient art at the Cleveland Museum of Art for 28 years from 1969–1997. She has written two books about Amenhotep III, *Egypt's Dazzling Sun Amenhotep III and His World* (1993) and *Amenhotep III: Egypt's Radiant Pharaoh* (2011). She proposes that some sort of infectious disease epidemic, most likely bubonic plague, explains the 8-year hiatus in the recording in Amenhotep's reign, which she outlined in the article "Bubonic Plague During the Reign of Amenhotep III?" A summary of Kozloff's evidence presented in that article follows.

Image 3.3 Statue of Amenhotep III (© Gemma Ivern/iStock/Thinkstock)

There was the significantly increased interest in the Goddess Sekhmet. She was the Goddess of War and Pestilence and was regarded as "a relatively minor deity".[6]

Even during times of increased war activity, as occurred during the reign of Thutmose III who reigned from 1479–1425 BCE and was called the "Napoleon of Egypt" because he expanded Egypt's empire into the Levant, there was no significant increase in the worship of Sekhmet as the Goddess of War. But during Amenhotep's reign, when he had no significant military action, he "commissioned more monumental statues of Sekhmet than of all the other gods put together".[7]

Amenhotep III had over 1000 large statues of himself made during his reign and made less than 200 large statues of all other deities. But for the lioness-headed deity Sekhmet (see Image 3.4) he commissioned over 700 large statues—was this because he wished to evoke her powers to protect Egypt from a plague in her role of the Goddess of Pestilence?

The recording of bad news such as a plague by the ancient Egyptians was rare. There is little written about a plague during Amenhotep's reign, but there is an Amarna letter (EAII)[8] written during the time of Amenhotep's son Akhenaten's reign that mentions the plague during Amenhotep's reign. In the letter from King of Babylon Burnaburiyas to Akhenaten, Burnaburiyas was replying to Akhenaten's earlier letter in which he stated that one of his father's wives (but not his first wife Tiye) had died of plague. If a pharaoh's wife can die of the plague, so can the rest of Egypt.

Many scholars have examined the effects of infectious epidemics on the culture of the survivors. Millard Meiss looked at how the Black Death had affected painting in Florence and Siena in his book *Painting in Florence and Siena after the Black Death* published in 1951. A.A. Smith discussed the cultural and geographic effects of several plagues in his 1996 article "Plague in Ancient World: A Study from Thucydides to Justinian", while John Julius Norwich looked at the effect of the bubonic plague on the Byzantium Empire in his book *Byzantium: The Early Centuries* in 1988. Then there is the classic description of the Plague of Athens written by the Greek historian Thucydides (c. 460–c. 395 BCE) who lived through the epidemic and somehow survived the plague that killed most of the citizens of Athens.

Several common traits of how plagues affected the cultural and geographic characteristics of the survivors became apparent according to Kozloff[9]:

Image 3.4 Image of Sekhmet (© AmandaLewis/iStock/Thinkstock)

1. The plagues and epidemics cited occurred during periods of increased international trade and/or military movement.
2. Plagues, in particular bubonic plague, ravaged dense populations. This included monasteries or priestly groups, military and other highly concentrated populations, such as artisans, causing noticeable changes in religious and military activity and in artistic output and styles.
3. Panic caused flight from infected sites to "clean" areas.
4. Mass burials and other non-traditional methods of burial were practiced among diminished populations needing to dispose of large numbers of corpses. Mass graves where there is no evidence of foul play provide circumstantial evidence for the occurrence of mass death by disease, even though most infections leave no specific trace upon the bones.
5. In religious life, severe plagues caused disaffection for old beliefs and support for new belief systems. In some cases, this meant the death of prominent cults and the rise of new or minor cults.
6. Artistic representations changed both in style and subject matter after plague episodes. In some cases entire artistic colonies died out along with their artistic production. Where art continued to be made, it became more ritualistic and spiritual.
7. After plagues, marriages increased as part of efforts to replenish the population.

Kozloff then goes on to describe similar characteristics as listed above during the reign of Amenhotep III.

Characteristic 1: Plagues Usually Occurred in Times of Increased International Activity

Amenhotep's reign occurred during a time of great international trade in goods and people. Egypt engaged in trade and commerce not only with Mediterranean states but also in the Levant, the Middle East and the Indian subcontinent from where it will be shown later that the bubonic plague may have originated.

Characteristic 2: Devastation of Dense Populations

T. Säve-Söderbergh was part of the Scandinavian Joint Expedition to Nubia in 1967–1968. He wrote "Everything indicates a decreasing population from the Middle of the XVIII Dynasty and finally … (Lower) Nubia seems to be more or less depopulated".[10]

He also wrote that the tombs from the reign of Amenhotep III onwards were impoverished due to the lack of skilled stonemasons to make them and sculpt them. The Amarna letters also record that there was a lack of gold being produced presumably due to a lack of skilled workers. These workers were housed together in fortified settlements along the Nile. Such close knit population concentrations were perfect conditions in which a plague could spread.

The Black Death was a devastating infectious epidemic, traditionally thought to be caused by bubonic plague, which occurred in Europe in the mid-fourteenth century mainly from 1348–1350. It was called the Black Death because the skin of those infected with it turned black before they died. It occurred at the beginning of the Hundred Years War between England and France (1337–1453). During the period of the Black Death truces of several years duration were common. "It was no wonder that truces were concluded, which lasted for over two years; and were reviewed—despite some local bickering and raids by both sides—till 1354".[11] The same may have applied to Amenhotep's reign as both he and his enemies would have had depleted populations, too small to raise large armies to fight large-scale wars. That could be why Amenhotep's reign was relatively peaceful.

Characteristic 3: Flight to Clear Plague-free Areas

Amenhotep relocated his palace from Karnak to Malkata, this may have been provoked by a desire to find a plague-free area in which to live. Malkata was a mud-brick complex located on the West Bank of the Nile near Thebes, which is on the East Bank of the Nile and part of the modern-day city of Luxor. The Karnak Temple complex is located 2.5 km to the north of Luxor and is dedicated to the God Amun.

Malkata was Amenhotep's residence throughout most of the later part of his reign. "It is thought that Amenhotep III began construction of the Malkata Palace during the eleventh year of his reign … made this palace on the West bank the administrative center of his kingdom as well as his home. He housed his extensive harem in quarters in the palace complex".[12]

Note that construction started around year eleven of his reign which is the same time the 8-year "gap", possibly due to the plague, commenced – so it may have been built to flee the plague.

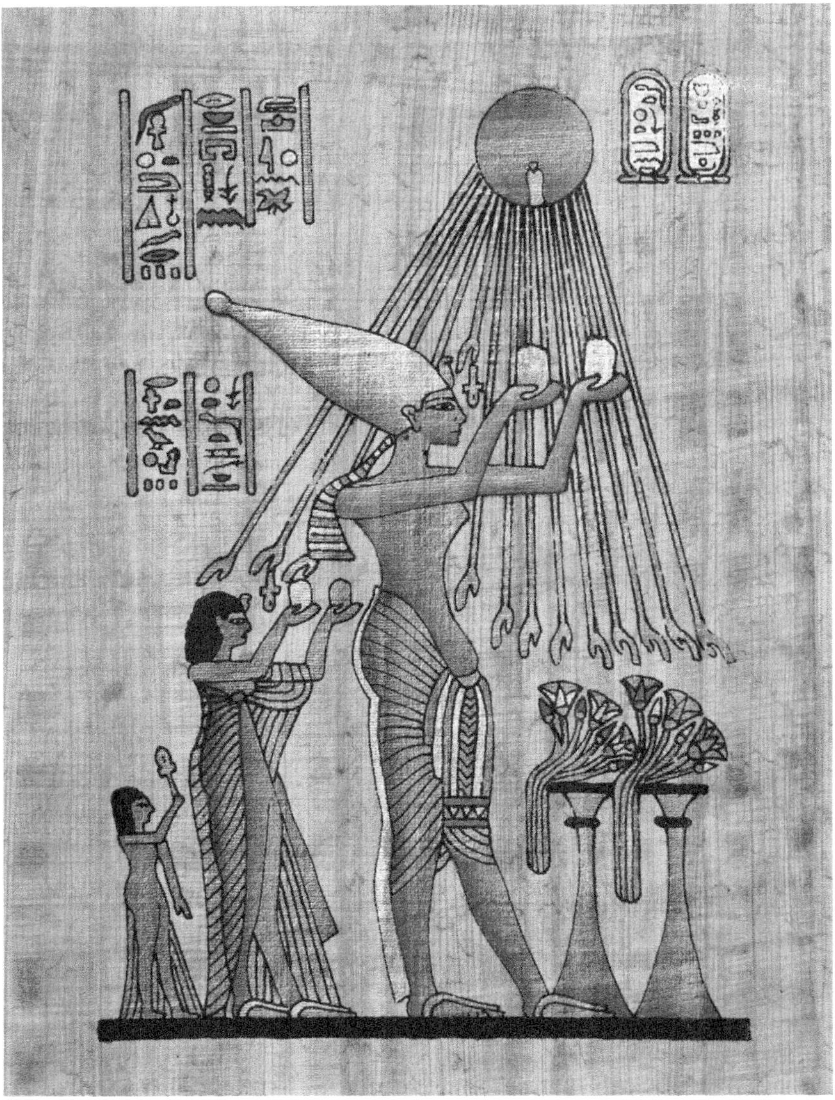

Image 3.5 Image of Akhenaten and his one God Aten (© Jose Ignacio Soto/iStock/Thinkstock)

The Egyptologist C. Aldred "felt that Akhenaten's move to Akhetaten was an attempt to find an area free of plague".[13] Akhetaten meaning "horizon of the Aten" (Aten being Akhenaten's new sole God) is also known as Amarna and was established by Akhenaten in c.1348 BCE (see Image 3.5). It was the usual practice of pharaohs to visit towns and travel from religious festival to religious festival along the Nile but Akhenaten did not travel and stayed only within the boundaries of his new city—it is possible that this was to protect himself from the plague.

Characteristic 4: Mass Burials, Non-traditional Burial Methods and Poverty of Grave Goods

Burials during Amenhotep's reign were unusual for three reasons: first there were more mass and multiple burials than before; second the burial chambers were poorly decorated and the offering objects were of poor and lesser quantity; and third there was an unusually high number of married couple burials, as if they had died nearly simultaneously.

Amenhotep's reign occurred in the time of "tremendous luxury and unprecedented wealth"[14] so it seems strange that most mass or multiple burials occurred in roughly cut and poorly decorated tombs. Was this due to a lack of skilled stonemasons and artists, who would have made the tombs in the usual Egyptian way—well cut and well painted? Was this lack of artisans due to their death in the plague?

A.H. Rind discovered an undecorated Theban tomb chamber in the mid-nineteenth century. It was sealed with Amenhotep's name and dated year twenty-seven of his reign.[15] The chamber contained no bodies but had artefacts for the king's sister Tiaa, his granddaughter Nebetta, many other princesses, butlers, guards, accountants and an embalmer.

One explanation for the tomb being empty is that the bodies had "been disposed of in a more expedient manner"[16] earlier because they had died of plague, and the empty tomb was a reburial cenotaph completed later. Rind also had "been surprised by the poverty of goods placed within the royal cenotaph. Poverty of grave goods is typical of plague".[17] This is as a result of the lack of artisans available to produce them.

Examples of married couples buried together include the king's parents-in-law Yuya and Thuya buried in the Valley of the Kings in an undecorated tomb, Henutwedjebu who was buried with her husband Hatiay in a rough cave and Kha and his wife. All tombs were rough and either undecorated or poorly decorated suggesting either that there was no time to plan for the burial or that there were too few artists available. Both scenarios point to plague.

Characteristic 5: New Cults and Changes in Religious Affiliation

The worship of gods associated with health, disease and healing increased during Amenhotep's reign. The most important of these was the increase in the number of larger-than-life statues of Sekhmet the goddess of war and pestilence. Her consort Ptah who was both punisher and healer also became more popular. His temple at Mennufer (Memphis) was expanded and redecorated during Amenhotep's reign at great expense. "The little dwarf-god Bes, protector of health and home and warrant against disease, also gained new levels of attention during the reign of Amenhotep III".[18]

Khonsu, an ancient god of the Old Kingdom who was famous for being blood thirsty, suddenly became a healing deity by the end of the New Kingdom. Amenhotep also made himself a god just before the Sed festival to celebrate the thirtieth year of his reign. Was this an extremely rare act to help him ward off the plague?

At Karnak there is a scribe statue that records how Amenhotep, son of Hapu, the king's most trusted and powerful official, had been ordered by the king to make a census of all the remaining Amen priests, and to fill their empty ranks after something (plague) had occurred throughout the Nile Valley.

The final piece of religious evidence favoring a plague is that Amenhotep's son Akhenaten instigated a religious revolution when he started monotheism with the God Aten. He may have done this because the traditional gods had failed him and he was trying desperately to save his kingdom from the plague. Also, if the numbers of traditional priests were greatly diminished due to dying from the plague, then there would be much less opposition to starting his new religion. So the old gods such as Amun were "out" and the new one god Aten was "in".

Characteristic 6: Changes in Artistic Subject, Style and Quality

Art during Amenhotep's reign became more spiritual and ritualistic. Early in his reign tomb painting flourished but in the middle of his reign tomb painting nearly ceased. Was this due to a lack of artists?

> Instead there seems to be an increase in painted wooden boxes and chests, executed by artists of lesser skill than the earlier tomb painters The decrease in skill level of artistry is typical of plague times; it suggests that the tomb painters died out, or at least declined in numbers, several years before the end of Amenhotep III's reign.[19]

The tomb painters lived at Deir el Medina, which was destroyed by fire during the time of Amenhotep or Akhenaten. If Deir el Medina was infested with plague, which was killing the tomb painters, then the best way to cleanse the site would be to destroy it with fire.

The subject matter of the paintings also changed. During the early part of Amenhotep's reign tomb painting featured traditional happy scenes such as fishing, fowling and large banquets with extended family. Later in his reign this changed to more serious and spiritualistic images with images of the king, limited family members and mourning scenes. There were also fewer animals depicted because they may have been banished for fear they caused the disease. "The highly specialized glass-vessel industry also declined towards the end of the reign of Amenhotep III. It died out completely during the reign of Akhenaten, and was not revived until early Ramesside times".[20]

Kozloff provides a further example of the deterioration of artisan standards, this time in sculptural masons: After an eight-year total lapse in the written record of Amenhotep III's reign, the first text to appear is the year 20 record of the promotion of a temple official, Nebnefer, which occurred in the presence of four high priests of Amen. This extremely important document is not inscribed on a large and expensive granodiorite stela or statue, but rather around the sides of a small, poorly executed limestone statuette, giving the impression that the best hard-stone sculptors and engravers were not available.[21]

All the changes point towards a major loss of artistic skill and talent during the time of Amenhotep and Akhenaten. As there were no major wars or famines to account for the deaths the disease is a prime reason.

Characteristic 7: Increase in Marriage

Incest was avoided by the ancient Egyptians and it was very rare for a pharaoh to marry his daughter, so when Amenhotep III married not only one daughter but two, namely Princess Sitamen and Princess Iset both by his first wife the Great Queen Tiye, most scholars assumed "that these marriages were merely symbolic or ceremonial".[22] Amenhotep performed the ceremony just before his first Sed festival in year thirty of his reign and may have had another reason for marrying his daughters—to replenish the numbers of the royal family with pure stock, because its members had been killed off by the plague.

Amenhotep had also made himself a God to perhaps make it more acceptable to marry his daughters who were of child bearing age, instead of marrying his sisters who were too old to become pregnant. So instead of the marriages being pure lust and incest they were transformed to a higher level that a God doing his divine duty of ensuring the survival of his kingdom and royal family after a disaster, such as a plague, had killed so many.

Amenhotep did not stop there:

> By year 30 Amenhotep was "either bargaining for an already married to (his year 10 Mitannian wife Gilukhipa's) niece Tadukhipa, two Babylonian princesses and the daughter of the King of Arzawa". These foreign marriages cannot be attributed entirely to plague response because such alliances had been joined before the Eighteenth Dynasty; however the number of Amenhotep's foreign queens far exceeded the norm.[23]

Also in year thirty of his reign Amenhotep made many women exempt from tax so that they would become financially secure enough to marry men who otherwise would not have been able to support a wife and children. This act by Amenhotep was a marriage incentive to help repopulate the empire. The women exempted from taxes included "hairum girls and songstresses for the Amen estate, as well as 'free women who were servants since the time of his (Amenhotep III's) forebearers'".[24]

Akhenaten continued the incestuous marriage practice of his father. He married a sister and with her fathered Tutankhamun (the pharaoh made famous by his intact tomb's discovery in 1922 by Howard Carter). He also married three of his daughters namely his eldest daughter Meritaten, his second daughter Meketaten and his third daughter Ankhesenpaaten who in turn would later also marry her half-brother Tutankhamun. Were all these incestuous relationships also in response to the plague and the need to restock a depleted royal family with genetically "pure" stock?

In summary, following a period of trade with Western Asia, Egypt was subject to an 8-year "gap" period of no written record between the years of twelve to twenty of Amenhotep III's reign. This was followed by the commissioning of many hundreds of over-life-sized hard-stone statues of the God of Pestilence Sekhmet; which would have left fewer artisans available to do other work, in addition to those that had already died due to the plague.

It also caused Amenhotep to move from his royal palace in Karnak to his new clean area site in the middle of the desert at Malkata, because he may have observed that green areas such as Karnak on the banks of the Nile (where rats would have flourished) suffered badly from the plague, whereas desert areas such as Malkata (where rats would not have flourished as well) were less likely to suffer from the plague.

Because so many infants and small children had died during this first bout of the plague it meant that by year thirty of Amenhotep's reign there was an acute shortage of women of child-bearing age. In response Amenhotep exempted, from paying taxes, the three classes of women who would have survived the first bout of the plague (see above). Children, who survived the plague, were now reaching child-bearing years so they could marry and have children.

The second bout of the plague may have occurred around year twenty-seven, which would explain the unusual mass burial tomb found by Rind and also explain why Amenhotep made himself a God and married his two daughters just before his year thirty celebrations. Another bout may have occurred in the early years of Amenhotep III's son's reign. In response to this Amenhotep IV changed his name to Akhenaten, changed Egypt's religion to Atenism because all the old Gods had failed to protect him and Egypt from the plague, and moved his royal palace to his own clean area site at Amarna.

So what could have caused the original 8-year gap in recording and subsequent episodes of a lack of recording? As stated earlier Amenhotep did not fight any major wars that would have killed off large numbers of young men but not women, children and artisans. Nor was Egypt invaded during his reign, which could have resulted in many men, women, children and artisans being killed. There were no major famines either. All evidence points to an epidemic or epidemics by some sort of infectious plague to account for such loss of life amongst men, women and children and the lack of skilled artisans.

Research conducted by Prof. Eva Panagiotakopulu will answer the question of which type of plague. There is evidence that suggests trade between Pharaonic Egypt and India occurred: Panagiotakopulu found the Khapra beetle, a native of India, in wheat from a Middle Kingdom tomb at El

Gelelein.[25] If the trade between ancient Egypt and ancient India introduced bubonic plague into ancient Egypt then that same bubonic plague must have existed at the source, that is, ancient India. This bubonic plague may then have contributed to the demise of the Indus Valley Civilization as well.

Yersinia (*Pasteurella*) *pestis* is the name of the bacillus that causes the bubonic plague. It can be transmitted to humans via various fleas but classically via the black rat flea *Xenopsylla cheopis*, which can also live on a number of other animals such as cats, guinea pigs, dogs and carnivorous animals that eat rats. Other fleas that could carry the plague bacillus include the human flea *Pulex irritans* and the cat flea *Ctenocephalides felis*. Egyptians were very fond of cats and actually worshipped them. Panagiotakopulu has researched at the Workman's Village at Amarna, Egypt and found human fleas there along with cat fleas.[26] Note that this is the earliest known discovery of the human flea in the Old World.

Boessneck found black rat remains at his dig site of Tell Eldab'a in the Nile Delta dating from 1700 to 1600 BCE, along with Nile rat remains.[27] Plague like illnesses have been mentioned in Egyptian papyri as the following quote by Dr. E. Panagiotakopulu shows:

> Perhaps the first time plague is mentioned is in the Ebers Papyrus, a medical papyrus dated around 1500 BC, but probably compiled much earlier:
>
> If thou examines a man who suffers from the said (i.e. from the shivering fit described in 38.3-10) for hours, like consuming for purulency, and he is weak like a breath that passes away, then thou shalt say that it is (due to) closing (?) of an accumulation, which cannot be raised and does not trust in a weak remedy; it (i.e. the accumulation) has produced a bubo, and the pus has petrified, the disease has hit. Thou shalt prepare him remedies to open it by means of medicines. (Ebers Papyrus 39, translated by Ebbell, 1937).

Another tentative description of a disease that could be the plague is found in the Hearst Medical Papyrus, a text paleographically related to the Ebers Papyrus, and dated to c.1520 BCE: one of the incantations is against "the Canaanite illness":

> Who is knowledgeable like Ra? Who knows the like of this God? – When the body is blackened with black spots – to arrest the God who is above. Just as Seth had banned the Mediterranean Sea, Seth will ban you likewise. O Canaanite illness! You shall not intend to pass through the limbs of X, born of Y. (Hearst Papyrus H XI 12–15 translated by Goedicke, 1984, p.94).[28]

These quotes from the papyri mention "bubo ...the disease has hit" and "...the body is blackened with black spots..." Both these descriptions are consistent with infection by bubonic plague. Dr. Panagiotakopulu continues her quote:

> "In the London Medical papyrus, dated 1350 BC, there are incantations 'in the language of Keftiu' against the 'Canaanite illness' (Goedicke, 1984). A further possible description of the disease could be in the incomplete section of the same papyrus"
>
> When the body is coal black with charcoal (spots) in addition to the water (urine) as red liquid (i.e. bloody).... (London Medical Papyrus 15, 8–10, translated by Goedicke, 1984).

Goedicke (1984) is definite that this disease is bubonic plague, although the evidence is insubstantial. This disease was probably called the Canaanite or Asiatic (Amu) illness, either because of the place in which it originated, or simply because everything deleterious tends to be blamed on somebody else. The case of syphilis, Shakespeare's 'Malady of France' (Henry V, Act V, Scene 11), provides a more modern example of the same.

One of the Amarna letters, a collection of incoming diplomatic correspondence sometimes very exaggerated, also refers to plague. In letter EA 35, the king of Alasia (? Cyprus) explains why the amount of copper he sent to the king of Egypt is small:

> Behold the hand of Nergel (pestilence, plague) is now in my country; he has slain all the men of my country, and there is not a (single) copper-worker. (Translated by Moran, 1992)[29]

It can be seen then that there must have been a devastating infectious disease in Alasia, so bad that it killed off all the men. Thus there would be no men to grow food leading to famine, no men to mine minerals so less trade and income for the country, no men to administer the country, let alone defend it from invasion leading to its possible demise.

Panagiotakopulu has proposed the following hypothesis—refer to the chart in her article (see note 28):

> Amarna had a short history being abandoned after about 20 to 25 years. Akhenaten reigned from c1353–1334 BCE, and had Amarna constructed between years 5 and 9 of his reign. He used mud brick and whitewash,

> instead of stone, to speed up construction while important buildings were faced in local stone. It was abandoned after the death of Akhenaten. Was this due to no more patronage by the pharaoh or was it because plague infected the city as "the sudden deaths of several members of the royal family could be linked to the plague"?[30]

It can now be seen that bubonic plague existed in Egypt at the time of Amenhotep III and Akhenaten, if not earlier. Bubonic plague also recurs in epidemics, many years apart, when circumstances permit. So it is logical to think that the infectious disease introduced into the Hittite Empire by the Egyptian prisoners of war in 1322 BCE was possibly a recurrence of bubonic plague, because there is evidence of it in Amarna only a few years earlier from c.1348–1334 BCE and from c.1378 BCE (after year twelve of Amenhotep's reign).

That is why German archaeologists refer to the Hittite Epidemic of 1322 BCE as an outbreak of bubonic plague "another disaster will punish his (Suppiluliumas') Empire: the plague, more precisely, the bubonic plague"[31] or in German "die pest, genauer: die Beulenpest"[32] rather than smallpox as the earlier doctors had thought. This bubonic plague theory for the cause of the 1322 BCE Hittite Epidemic is also supported by Eric Watson-Williams[33] and Lars Walloe.[34]

The end of the Hittite Empire occurred over 120 years later in c.1200 when it was part of the bigger catastrophe known as the end of the Bronze Age.

Notes

1. Gardiner, A., *The Egyptians*, (London: Folio Society, 2005), 433. Other dates for his reign include June 1386 to 1349 BCE or June 1388 to December 1351 according to Jurgen van Beckerath in *Chronologie des Pharaonischen Agypten* (Philip von Zabern Mainz, 1997) 190.
2. Ibid.
3. Gardiner, A., (op.cit) 200.
4. O'Connor, O and Cline, E., (1998), *Amenhotep III: Perspectives on His Reign*, (University of Michigan Press, 1998) 11–12.
5. Kozloff, A.P., "*Bubonic Plague During the Reign of Amenhotep III?*" (KMT, Vol 17, Number 3, 2006) 36.
6. Ibid and Sadek, A.I., *Popular Religion in Egypt during the New Kingdom*, (Hildesheim, 1987) 29. He noted that her cult arose in the New Kingdom, becoming especially strong during the reign of Amenhotep III.

7. Ibid.

Note for Reference 7: This is also verified by the St. Louis University course on Egyptology -http://euler.slu.edu/~bart/egyptianhtml/kings%20and%20Queens/amenhotepiii.htm, 5. Accessed 7 December 2012 where it states "At some point Amenhotep III had many statues of Sakhmet erected in the temple of Mut precinct in Karnak. Many hundreds of these statues have been found over the years. Some have speculated that the protective nature of the goddess may point to an attempt to have the gods protect Egypt against disease".

Also refer to—Sekhmet, http://www.egyptianmyths.net/sekhmet.htm, 1. Accessed 9 December 2012 that states "Having once unleashed her powers for the destruction of mankind, the Egyptians feared a repeat performance by Sekhmet. The Egyptian people developed an elaborate ritual in hopes she could be appeased. This ritual revolved around more than 700 statues of the goddess. The ancient Egyptian priests were required to perform a ritual before a different one of these statues each morning and each afternoon of every day of every single year. Only by the strictest adherence to this never-ending ritual could the ancient Egyptians be assured of their ability to placate Sekhmet".

Also refer to—Sekhmet, http://www.pantheon.org/articles/s/sakhmet.html 1. Accessed 9 December 2012 that states "She [Sekhmet] was also the 'Lady of Pestilence' who could send plague and disease".

Also refer to—Sekhmet, http://ancientegyptonline.co.uk/Sekhmet.html 1. Accessed 9 December 2012 that states "Amenhotep III (father of Akhenaten, Dynasty Eighteen) built hundreds of statues of Sekhmet in the precinct of Mut's temple (known as "Isheru") south of the Great Temple of Amun in Karnak. It is thought that there was one for every day of the year and that offerings were made every day".

8. Moran, W.L., *The Amarna Letters*, (Baltimore, 1992) 21.
9. Kozloff, (op.cit.) 40.
10. Säve-Söderbergh, T., "*Preliminary Report of the Joint Scandinavian Joint Expedition*", (Kirsh IS, 1967-68) 237–240.
11. Oman, C.W.C., "*The Hundred Years War*", (Chicago: Encyclopedia Britannica XI, 1943), 888b.
12. Malkata Palace, http://www.ancientegyptonline.co.uk/malkatapalace.html, 2. Accessed 3 December 2013.
13. Kozloff, (op.cit.) 42 and Alfred, C., *Akhenaten: King of Egypt*, (London, 1998) 149 and 283.
14. Ibid.
15. Dodson, A. and Janssen, J.J., "*A Thebian Tomb and its Tenants*", (Journal of Egyptian Archaeology, 75, 1989) 125–138.
16. Kozloff, (op.cit.) 42.

17. Ibid.
18. Ibid, 43.
19. Ibid, 44.
20. Ibid.
21. Ibid, 41 and Davies, B.G., "*Egyptian Historical Records of the later Eighteenth Dynasty*" (V, Warminster, 1994) 44–45.
22. Ibid, 44.
23. Ibid and University College of London course on Egyptology at— http://www.digitalegypt.ucl.ac.uk/ideology/sed/index.html, 1. Accessed 9 December 2012 which states, when discussing the Sed festival, "The purpose of the festival seems to have been the renewal of the physical and supernatural energies of the king".
24. Ibid.
25. Panagiotakopulu, E., "*Insect Remains from the Collections in the Egyptian Museum in Turin*", (Archaeometry, Vol. 45, Number 2, Wiley-Blackwell, 2003) 355–362.
26. Panagiotakopulu, E., "*Fleas from Pharaonic Amarna*", (Antiquity 75, 2001) 499–500.
27. Boessneck, J., "*Tell El Dab'a III Die Tierknochenfunde 1966-1969*", (Untersuchungen Des Osterreich Archaologischen Institut (UZK), Abt.S.Akademie Der Wissenschaft, Cairo, 1976.
28. Panagiotakopulu, E., "*Pharaonic Egypt and the origins of Plague*", (Journal of Biogeography 31, 2004) 272–273.
29. Ibid and Armitage, P.L., "*Unwelcome companions: ancient rats reviewed*", (Antiquity, Vol 68, 1994) 232.
30. BBC-History-Ancient History in depth: "*The End of the Amarna Period*", 5.
31. The Hittites, http://www.michaelmaxwolf.de/antike/alter_orient/hethiter.htm, "*The Hittites*", 7 Accessed 3 December 2013.
32. Watson-Williams, E., (op.cit.).
33. Ibid, 109–125.
34. Walloe, L., (op.cit.).

CHAPTER 4

How Disease Affected the History of the Hittite Empire

A Background

The Hittite Empire lasted from approximately 1700 to 1200 BCE and was located in central Anatolia (modern day Turkey), with its capital called Hattusa located near the current Turkish city of Bogazkoy (see Image 4.1). At its height during the fourteenth century, the Hittite Empire extended from central Anatolia into southwestern Syria as far as the city of Ugarit where it bordered the northern end of the Egyptian Empire and into upper Mesopotamia where it bordered the Assyrian Empire. After its collapse around 1200 BCE the Hittite Empire disintegrated into several independent "Neo- Hittite" city-states.

The Hittites spoke an Indo-European language and may have originated from the Pontic Steppe in modern Ukraine around the Sea of Azov in the third and fourth Millennia. They moved southwest around the Caspian Sea around 1900 BCE until they came to central Anatolia, which was then occupied by the Hattian people and some Assyrians. The Hittites eventually adopted the Assyrians' cuneiform tablet scripts themselves and it is from these tablets (over 24,000 tablets alone have been found from the archaeological dig at Hattusa), plus diplomatic and commercial communications found in Egyptian and other Near Eastern archaeological sites, that we can make up a history of the Hittite Empire; because very little of it remained after its demise to tell its story.

P. Norrie, *A History of Disease in Ancient Times*,
DOI 10.1007/978-3-319-28937-3_4

Image 4.1 Photograph of The Lion's Gate Entrance to the Hittite Capital Hattusa (© ImageCaravan/iStock/Thinkstock)

The Hittite name is a modern name derived from the Hattian people who spoke a non-Indo-European language that modern anatolianists call "Hattic". The Hittite Empire is divided into the Old Kingdom starting with its first king Labarna I prior to 1650 BCE and ending in 1400 BCE with the reign of Muwatalli I. The New Kingdom started with Tudhaliya I in 1400 BCE and ended with Suppiluliuma II who came to the throne in 1207 BCE and oversaw the final collapse of the Hittite Empire in about 1200 BCE.

Hattusili I came from his capital in Hattusa and conquered the areas to the north and south before campaigning to the southeast into the Kingdom of Yamkhad in Syria. Hattusili's son Mursili I consolidated his father's gains in Syria by capturing the city of Aleppo, the capital of the Kingdom of Yamkhad. He then attacked Mesopotamia and eventually captured its capital Babylon in 1531 BCE. These lengthy campaigns strained the Hittite Empire's resources and its capital Hattusa.

After Mursili I returned to Hattusa he was assassinated so the Hittite Empire was plunged into chaos and the first of many power struggles for its control. Throughout its history the Hittite Empire was plagued by infighting within the extended Hittite royal family and upper echelons of Hittite society for control of the Hittite Empire, sometimes resulting in political assassinations and civil war. Because of the intrigues, the Hittite Empire was sometimes weakened and vulnerable so the rulers had to devise ways to protect their borders from invasion. Thus the Hittites became pioneers in the art of international politics and diplomacy. They conducted treaties and alliances with their neighboring states, a great innovation by their rulers to prevent war and guarantee peace.

After the murder of Mursili, the Hittite Empire was weakened so the Hurrians took advantage of its vulnerable state. The Hurrian people came from the mountainous region along the upper reaches of the Tigris and Euphrates Rivers. They captured Aleppo and the coastal region of Adaniya and renamed it Kizzuwatna. Later the Hittite King Telelpinu allied himself with this new Hurrian state of Kizzuwatna against another Hurrian State, the state of Mitanni in southwestern Syria, thus once again renewing Hittite expansion into Syria. Later this expansion continued under King Tudhaliya I who again allied himself with Kizzuwatna to fully conquer the other Hurrian states of Aleppo and Mitanni and expand west into Arzawa.

After the reign of Tudhaliya I there followed another weak phase where invaders destroyed the Hittite capital Hattusa, but its power and prestige was restored under the reign of Suppiluliuma I (c1350–1322 BCE). He was a dynamic king who conquered several Syrian city-states such as Aleppo, Mitanni and Carchemish, putting his sons in charge of the new conquests. This made the new resurgent Hittite Empire the only rival to Egypt in the region because Babylonia was still under Kassite rule and Assyria was only newly independent after the Hittite conquest of the Mittani.

Suppiluliuma I's reign ended when he died during the 1322 BCE epidemic along with his eldest son Arnuwanda II leaving the kingdom to his second son, Mursili II. The kingdom Mursili inherited was strong in the east in Syria but weak in the west of Anatolia, so he attached Arzawa and the city of Millawanda in the coastal land of Ahhiyawa. Mursili also inherited a kingdom being devastated by the 20-year epidemic that started in 1322 BCE.

The wealth of the Hittite Empire was born from its trade routes for metals through northern Syria into Mesopotamia. When these trade routes were threatened during the reign of Muwatalli II by Egyptian expansion northward during the reign of Ramesses II, a clash between the two regional superpowers was inevitable. This clash was the Battle of Kadesh, which occurred in Syria in c.1275 BCE.

Image 4.2 Photograph of Hittite Greenstone Cube (© mathess/iStock/Thinkstock)

At this time the Assyrians were extending their kingdom by seizing the Kingdom of Mitanni and expanding to the Euphrates River. In order to protect its borders with Egypt and Assyria, the Hittites under Hattusili III signed a peace treaty with Egypt still under Ramesses II in c.1258 BCE. The "Treaty of Kadesh" is the oldest written surviving peace treaty in the world and a copy of it hangs in the United Nations Assembly Building in New York, USA.

Hattusili's son Tudhaliya IV was the last strong Hittite king. He kept the Assyrians out of Syria and even temporarily annexed the island of Cyprus for his empire, but unfortunately he could not prevent the Hittite Empire's final demise.

The last Hittite king Suppiluliuma II came to the throne in 1207 BCE. He won a naval battle against the "Sea Peoples" off the coast of Cyprus, but this was a late gesture of defiance as the "Sea Peoples" continued to conquer the Near East lands, cutting off the Hittite trade routes and isolating the Hittite Kingdom, which eventually ended in the early twelfth century BCE. For a list of Hittite Kingdoms and Kings refer to Gurney's *The Hittites.*[1]

Did the Hittite Empire End with the 1322 BCE Epidemic?

Introduction

This section will discuss the 20-year long epidemic that infected the Hittite Empire, which started in 1322 BCE. It will discuss possible causes and critically examine the evidence of the 1322 epidemic.

Three eminent medical doctors and historians have suggested that the 1322 BCE Hittite epidemic was so bad that it ended the Hittite Empire. The three are William Wynn (Professor of Medicine at the University of Birmingham Medical Faculty), Ronald Hare (Professor of Bacteriology at the University of London) and M.P.M. Cooray MBBS (London), MRCS (Eng), LRCP (London) who was a former physician at the Fever Hospital in Angoda and President of the College of General Practice in Sri Lanka. In 1980, he wrote a book *The Story of Man and His Philosophy* that was published by the Anada Press in Colombo, Sri Lanka and the M.P.M. Cooray Memorial

Image 4.3 Photograph of Hittite Kings Tomb (© SafakOguz/iStock/Thinkstock)

Oration is given in his memory each year at the Faculty of Medicine at the University of Colombo.

However, the Hittite Empire did not end with the 1322 BCE epidemic, it lasted for another 120 years or more until the reign of Suppiluliuma II, who came to the throne in 1205 BCE. For a full list of Hittite Kings refer to "The Hittites" by O. R. Gurney.[1] The Hittite Empire also fought the battle of Kadesh in c.1275 BCE, that is, 27 years after the end of the 20-year long 1322 BCE epidemic. The battle of Kadesh was the first ever recorded battle in ancient times and it was between the two super powers of the time namely Egypt and the Hittite Empire, and it occurred where their spheres of influence clashed in Syria.

The Hittite Empire ended around 1200 BCE and was part of the catastrophe that was the end of the Bronze Age, which lasted from approximately 1200 BCE until 1150 BCE. The current consensus theory for the fall of the Hittite Empire was a combination of three main things: first famine; second rivalry between King Suppiluliuma II, the last Hittite king who came to the throne in 1205 BCE, and a relative leading to civil war; and third a general weakening of the central control of the empire so that distant vassal states began to revolt leading to final invasion by outsiders such as the Kaska people.

There is plenty of evidence about food shortages in Hattusa near the end of the thirteenth century. Ura was the port on the southern coast of Anatolia through which grain shipments were imported, mainly from Egypt and Syria. In the late thirteenth century, most likely during the reign of Tudhaliya IV (1237–1209), the vassal state of Tarhuntassa, which included the port city of Ura, was lost to the Hittite Empire, thus adversely affecting grain shipments to Hattusa. The pharaoh Merneptah referred to a shipment of grain which he had sent to "keep alive the land of Hatti" in one of his inscriptions at Karnak.[2]

The following letter from the Hittite court to the King of Ugarit demands a ship and crew to transport the equivalent of 450 tons of grain from Makish to Ura urgently, (note that "My Sun" is the reference given to the Hittite King).

> And so (the city) Ura (acted (?)) in such a way … and for My Sun the food they have saved. My Sun has shown them 2,000 kor of grain coming from Mukish [A "kor" was an ancient Hebrew unit of measurement equal to 6 liters in today's units]. You must furnish them with a large ship and crew, and they must transport this grain to their country. They will carry it in one or two shipments. You must not detain their ship![3]

Throughout the history of the Hittites rivalry between family members for the throne, especially between brothers and cousins was common. The following set of instructions from the Hittite King Tudhaliya to his staff and high officials, in which he demands their unconditional loyalty, shows how he feared that the various members of the extended Hittite royal families, such as his cousin Kunuata, wanted to take the throne from him in favor of their chosen family member.

> My Sun has many brothers and there are many sons of his father. The Land of Hatti is full of the royal line: in Hatti the descendants of Suppiluliuma, the descendants of Mursili, the descendants of Muwatalli, the descendants of Hattusili are numerous. With regard in kingship, you must acknowledge no other person (but me, Tudhaliya), and protect only the grandson and great grandson and descendants of Tudhaliya. And if at any time (?) evil is done to My Sun – (for) My Sun has many brothers—and someone approaches another person and speaks thus: "Whomever we select for we need not even be a son of our lord!"—These words must not be (permitted)! With regard to kingship, you must protect only My Sun and the descendants of My Sun. You must approach no other person.[4]

Similarly the reign of Suppiluliuma II was under constant threat from members of the extended royal family, which would have had "a destabilizing effect upon the monarchical structure and may well have seriously undermined the king's authority in the eyes of many of his subjects. This is reflected in the rebellion in Hatti and perhaps also in the insubordination of vassal rulers, both reported in Suppiluliuma's reign".[5]

Hittitologist Singer has commented on this aspect of Hittite history stating "Without diminishing the role of the outside enemies in the fall of the Hittite Empire, I feel that more weight should be given to the symptoms of inner decline and disintegration".[6] Internal political instability due to royal family rivalry could have led to civil war within the Hittite Empire.

The final current theory for the demise of the Hittite Empire was invasion from outside vassal states such as the Kaska people to the north of the land of the Hatti. The Hittite Empire and the state of Kaska had long been enemies, and with the capital Hattusa abandoned, the Kaska people may have invaded to deliver the final blow.

> Suppiluliuma II abandoned his capital Hattusa in about 1200 BCE. Seeher is the director of excavations at Hattusa and:

His scenario is one of gradual abandonment of the capital, firstly by the royal family and leading members of the palace bureaucracy who took with them all their valuable and portable possessions, including the kingdom's most important official records. They must have done so once it became clear that the capital was doomed. Certainly there is evidence that many buildings in the city were finally put to the torch, by this time, according to Seeher, the city had become largely derelict".[7]

Why was the city "doomed?" Was it because of an infectious disease epidemic killing everyone; hence the occupants of Hattusa had no alternative but to abandon their capital? Professor Bryce goes on with his explanation of Seeher's findings at the Hattusa excavations:

Those who stayed behind were left to fend for themselves as best they could, scavenging at leisure through the leavings of those who had departed. The decline, abandonment, and final destruction of Hattusa probably occurred in the very early years of the 12th century. The whole process may have taken no more than a few months".[7] Some parts of the city were burnt and "archaeological evidence indicates widespread devastation by fire in the capital—on the royal acropolis, in the temples of both Upper and Lower Cities, and along stretches of the fortifications. This has conjured up the scenario of a royal capital succumbing all at one time to violent destruction in an all-consuming conflagration.[8]

Trevor Bryce argues in his book *The Kingdom of the Hittites* that the Hittite Empire collapse should be viewed as part of a general Near Eastern collapse.

But in attempting to find reasons for the collapse of the Hittite kingdom, we should be careful not to give undue prominence to any specific set of factors, whether internal or external. Further, its collapse did not occur in isolation. The fact that a number of centers of the Mycenaean world were destroyed, in roughly the same period as the fall of Hatti and other Near Eastern kingdoms, gives some credence to the view of a series of widespread upheavals and disasters, at least within the Greek and Near Eastern worlds, which led to or helped precipitate the downfall of the major centers in both regions. Hence the theories of a long-lasting and ruinous drought, or of simultaneous or contemporaneously related onslaughts by "Sea Peoples" upon both the Greek and Near Eastern worlds, or of a widespread "systems collapse". While we should be mindful that there were significant differences in the patterns of decline and collapse of the Mycenaean and Near Eastern centers

> of power, it is difficult to believe that there is not some relationship between the course of events in both regions in the last decades of the thirteenth century and the early twelfth century.[9]

Hence we have theories such as volcanic eruptions, earthquakes, tsunamis, drought, famine, invasion by "Sea Peoples", a general systems collapse, changes in warfare and internal fighting within royal families trying to explain this general Near Eastern collapse. But what about disease being a cause as well? Epidemics of various infectious diseases could also cause Bryce's above quoted "a series of widespread upheavals and disasters". Up till now only Tom Slattery in his book *The Tragic End of the Bronze Age—a Virus makes History* plus Watson-Williams and Walloe have suggested disease could be a possible cause or contributor to the End of the Bronze Age. Slattery's book referred only to smallpox and the other two books referred only to bubonic plague, not other possible infectious causes. This book will examine other types of infectious diseases such as tularemia (or the Hittite Plague), tuberculosis, leprosy, influenza and polio, to name but a few, as well as examine the facts about the 1322 BCE epidemic in the Hittite Empire in detail for the first time.

The Epidemic of 1322 BCE in the Hittite Empire

In 1954, Ronald Hare published his book *Plague and Pestilence Infectious Disease. Its Origins and Conquest* in which he stated:

> It is also possible that it (smallpox) occurred amongst the Hittites, since cuneiform tablets found in Asia Minor record a severe epidemic which Wynn considers may have been smallpox (see embedded note vi). The epidemic in question is of some historical importance since most of the Hittite army and a large proportion of the civilian population, as well as the King and the son who succeeded him, died during it. This crippled them to such an extent that they were overwhelmed by the Achaeans and Assyrians, bringing to a sudden end a civilization which had flourished for a thousand years.[10]

The following is a profile of Hare from the Archives in London: "Hare was a bacteriologist of repute working through an exciting period in the history of the discipline. Among other activities he reported on the scientific value of bacteriological experiments undertaken in German concentration camps during World War Two (see PP/HAR/B.7). He was also an historian of the subject. Born 1899; Royal Masonic School,

Bushey, Herts, 1910–1917; Birkbeck College, 1918–1919; St. Mary's Hospital: LMSSA, MBBS, 1919–1924; Research scholarship, Institute of Pathology and Research, St. Mary's, 1925; Assistant, Inoculation dept., St. Mary's, 1926–1930, 1st Assistant, Research Laboratories, Queen Charlotte's Hospital, London, 1931–1936; Canada: Research Associate, Connaught Laboratories, University of Toronto, Lecturer in Department of Hygiene and Preventative Medicine; largely responsible for planning and building of Canadian Government penicillin plant at University of Toronto, 1936–1946; Professor of Bacteriology, University of London, 1946–1964; Honorary Consulting Bacteriologist, St. Thomas's Hospital, 1951; Member of Council of Wright-Fleming Institute, 1952–1960; Emeritus Professor of Bacteriology, University of London, 1964; died, 1986".[11]

Hare also knew fellow bacteriologist Sir Alexander Fleming and wrote a detailed history of Fleming's other less well-known activities called "The Scientific Activities of Alexander Fleming, other than the Discovery of Penicillin". This 25-page history was published in 1983 in the journal *Medical History*.[12]

Hare as a noted Professor of Bacteriology whose word would carry a lot of weight, stated in his book two things; first that the 1322 BCE Hittite epidemic was possibly smallpox and that it brought to a sudden end the Hittite Empire that had lasted over 500 years. Hare's statement was derived from the reference "6. Cited in annotation in *Lancet*, i, 69, 1949".[13] This reference refers to an annotation written in the *Lancet* of the 8 January 1949 about Fitzpatrick lectures given by W.H. Wynn at the Royal College of Physicians on the 7 and 9 December, 1948 titled "Plagues and the Fortunes of Men".

The Annotation stated:

> The earliest recorded epidemic befell the Hittites, a flourishing civilization in Mesopotamia from about 1750 BC onwards, though their culture went back much earlier than that. Cuneiform records on clay tablets found in Asia Minor tell how the son of the Hittite king set out to marry a widowed Queen of Egypt, but died soon after his arrival in her country. His angry father declared war on Egypt and invaded Syria; he won several victories, but sickness broke out among his soldiers and spread to the civil population on their return home. The disease (probably smallpox) remained endemic for twenty years and killed many–among them the king and his son who succeeded him. The Hittites were weakened so much that their subject

> tribes revolted, their army was crippled in the Egyptian war, and finally the Achaeans, after taking Troy, drove them into Syria where the Assyrians overwhelmed them. Their empire of a thousand years melted away leaving hardly a trace – heavy work for a virus.[14]

Hence this Annotation refers to the disease that caused the 1322 BCE epidemic that lasted 20 years and caused so much misery as "probably smallpox".

Wynn and the Fitzpatrick Lectures

The Fitzpatrick Lectures are lectures given at the Royal College of Physicians in London on "the History of Medicine" in memory of Thomas Fitzpatrick who was born in a small Irish country town called Virginia in 1832 and graduated with an MD from Trinity College in Dublin. He was a member of the Royal College of Physicians and became an Assistant Surgeon in the East India Company serving in Bengal. After an illness he returned to London to practice in Sussex Gardens. He was a noted linguist and "wrote about the social and sanitary conditions of the laboring classes in Ireland".[15]

Fitzpatrick died in 1900 and his widow, Mrs. Agnes Letitia Fitzpatrick gave the Royal College of Physicians £2000 in 1901 to found the Fitzpatrick Trust to endow the Fitzpatrick Lectureship, the first of which was given by J.F. Payne in 1903 with a lecture "On English Medicine in the Anglo-Saxon Period".

William Henry Wynn was born in Birmingham in 1878 and educated at the King Edward's Grammar School in Ashton. At the age of 16 he won a scholarship to Mason College, which later became the University of Birmingham. At age 18 he graduated with a BSc winning the gold medal in Physiology and in 1901 gained his MB with honors from London and MSc from Birmingham. In 1902 he gained his MD and worked at Queen's Hospital before being appointed pathologist at the General Hospital in Birmingham. At the age of 26, in 1904, he became an assistant physician. Later he became one of the great physicians of his time in England and was one of the first to fully appreciate the impact of bacteriology on clinical practice, and he himself was a first class bacteriologist and clinical pathologist. Many of his early papers were on the treatment of pneumonia and other infections and on prophylactic vaccinations.

In 1926, he became professor of medicine at the University of Birmingham, and he held the chair with distinction until 1943, when on leaving the active staff of the hospital he became professor emeritus.[16] His main interest was always pulmonary diseases and his book "The Problem of Consumption" was published in 1912. He was elected a Fellow of the Royal College of Physicians in 1922, and was an examiner from 1934 to 1938 as well as Censor of the College in 1944 and 1945. Wynn gave four Fitzpatrick lectures in a series called "The Pestilences of War". The first was given on the 7 December 1948 and called "The Early Civilization and Greece". This is the lecture that featured the Hittite Empire. The second lecture was given on the 9 December 1948 on "The Roman Republic and Empire". The third lecture given on the 6 December 1949 was "Sixth Century to Fifteenth Century", with the fourth and last lecture on the 8 December 1949 being "Sixteenth Century to Nineteenth Century".

All Wynn's papers are held by the University of Birmingham, where he was Professor of Medicine. Martin Killeen, the Rare Books Librarian at the Cadbury Research Library, Special Collections at the University of Birmingham, Edgbaston, Birmingham noted to the author of this book that:

> In our archives there is a box of the papers of W.H. Wynn which contains several typescript variant copies of parts of some of the lectures and about 85 pages of manuscript notes but unfortunately nothing which refers to small-pox and the Hittites. In reply to my email the Royal College of Physicians indicated that the note which refers to a manuscript copy at Birmingham relates to the third lecture only.[17]

The Royal College of Physicians does not have a copy of Wynn's first lecture nor does the Wellcome Collection in London. So there are no notes from Wynn's research about the 1322 BCE Hittite Epidemic to tell us why he thought it was "probably smallpox". It could have just been a calculated guess. Thus Wynn's "probably" became Hare's "possible" and eventually it became M.P.M. Cooray's definite "which medical historians now identify as smallpox".[18]

In 1965, M.P.M. Cooray wrote a paper "Epidemics in the Course of History", which was published in the *Ceylon Medical Journal.* In this paper he states "Cuneiform tablets found in Asia Minor record a severe epidemic which medical historians now identify as smallpox. The epidemic in question is of some historical importance since most of the Hittite army and a large population as well as the King and his son who succeeded him died in this epidemic. This crippled them that the Achaeans and Assyrians overwhelmed them bringing to a sudden end a civilization which flourished for a thousand years. This would have been about 1200 years BC".[19]

Both Hare and Cooray attribute the end of the Hittite Empire to the epidemic of 1322 BCE. But this is not so because the Hittite Empire lasted for another 122 years from the time of the beginning of the epidemic to the actual end of the Hittite Empire under King Suppiluliuma II in about 1200 BCE, and it fought the first recorded major battle of the ancient world, the battle of Kadesh (or Quadesh) between the Hittites and the Egyptians c1275 BCE. The Hittite Empire must have been still going strong in 1275 BCE, some 47 years after the beginning of the 1322 BCE epidemic, if it had to fight the might of the Egyptian Empire near the town of Kadesh on the Orontes River in southern Syria. Here a Hittite Army of some 50,000 men with 2500 war chariots under Muwatalli II fought an Egyptian Army of 20,000 men under Ramesses II. Both sides claimed victory but the Egyptians eventually retreated south and the Hittite Empire expanded southward to Upi.[20] If it had not been for the catastrophe and high death rate in the Hittite Empire during the 20 years following the 1322 BCE epidemic; could the Hittites have decisively won the Battle of Kadesh and then invaded Egypt itself?

Donald R. Hopkins in his Pulitzer Prize nominated book on the history of smallpox *The Greatest Killer, Smallpox in History*[21] has a more balanced view:

> The Hittite Empire was at its peak in the area of modern Turkey during the reign of Suppiluliumas I (c1380–46BC) when armies of the Hittite and Egyptian empires clashed several times in northern Syria. Cuneiform Hittite tablets describe one of these wars around the middle of the fourteenth century BC, after which the victorious Hittites were attacked by a pestilence. The disease originated among their Egyptian captives and spread to the Hittite army and civilian population, killing many ("Plagues and Fortunes", 1949). This epidemic raged among the Hittites for at least twenty years. Suppiluliumas I and his son Arnuwandas II, who succeeded him, both died of the disease within a year of each other. As a result, the early reign of inexperienced younger brother was marked by revolts and a series of invasions by neighboring states (Forrer 1937; Klegnel & Klegnel 1970). Order was eventually restored, and the Hittite Empire lasted another one and a half centuries before it was overcome by the Assyrians and Phrygians. Of the illness, unfortunately, we know only that it was contagious, apparently spread from person to person, and was often fatal. It could well have been smallpox, as both Cooray (1965) and Hare (1954) believe.[22]

According to Hopkins the Hittite epidemic "could well have been smallpox"[23] and is in agreement with this book that "the Hittite Empire lasted another one and a half centuries before it was overcome",[24] hence the 1322 BCE epidemic did not cause the sudden end of the Hittite Empire then.

Instead of repeating what has been said before by Wynn, Hare, Cooray and Hopkins, that smallpox was the most likely cause of the 1322 BCE Hittite Epidemic, this book will critically analyze the evidence to show another possible cause—bubonic plague. This book will also show later that outbreaks of bubonic plague reoccurred throughout the Near East and that it, along with tularemia and smallpox, could have contributed to the end of the Hittite Empire and the Bronze Age.

Donald Hopkins was born in 1941 and is an US physician, a MacArthur Fellow and is the Vice President and Director of Health Programs at The Carter Center. He graduated from Morehouse College with a BS, from the University of Chicago with a Doctor of Medicine, and from the Harvard School of Public Health with a Master of Public Health. He studied at the Institute of European Studies, at the University of Vienna.

During his career from 1984 to 1987, Hopkins was deputy director and acting director (1985) of the Centers of Disease Control and Prevention. Then he became an assistant professor of tropical public health at Harvard School of Public Health. He directed the Smallpox Eradication/Measles Control program in Sierra Leone in his capacity as a consultant to the World Health Organization. Hopkins "has received numerous awards, including the CDC Medal of Excellence, the Distinguished Service Medal of the U.S. Public Health Service, and a MacArthur Fellowship in 1995 for his leadership in the international campaign to eradicate Guinea worm disease".[25] His book, *Princess and Peasants: Smallpox in History* was nominated for the Pulitzer Prize in 1983.

Hopkins was also elected to the Institute of Medicine of the National Academy of Sciences in 1987 and has been a member of the American Society of Tropical Medicine and Hygiene since 1965. He was elected a Fellow of the American Academy of Arts and Sciences in 1997, awarded the Medal of Honor of Public Health (Gold) by the country of Niger in 2004, and named a Champion of Public Health by Tulane University in 2005. Hopkins currently serves on the Board of Directors for the MacArthur Foundation.

Hopkins has won many awards over the years: Centers of Disease Control Medal of Excellence, Distinguished Service Medal of the U.S. Public Health Service, 1995 MacArthur Fellows Program, James F.

and Sarah T. Fries Foundation Prize, Mectizan Award from Merck & Co., Knight of the National Order of Mali in 1998.

The Plague Prayers of Mursili II

What is the evidence for the 1322 BCE Hittite epidemic? To answer this question one must first look at the work of E.O. Forrer, an archaeologist and Hittitologist. In 1937, Forrer published a paper titled "The Hittites in Palestine, II".[26] In it he stated "The solution of this riddle, already given by me in 1926, is found in a prayer of Morsilis in the year 1331 BC, after the Egyptian pestilence had afflicted the Hittite empire for twenty years. The reason for this pestilence was held to be a sin of the Hittite king or state".[27]

The reference to 1926 was to Forrer's original translation of Hittite cuneiform tablets that appeared in his book *Forschungen II* (German meaning "Discovery") published in 1926. Ten fragments were translated and the translations appeared in pages 1–37 of his book.[28] These fragments contained the prayers to the gods written by Hittite King Morsilis (or Mursili II) asking for forgiveness and an end to the pestilence ravaging his country. These are now commonly referred to as the Plague Prayers of Mursili.

A second translation of the prayers was completed by fellow German archaeologist and Hittitologist Albrecht Goetze in 1930 titled "Die Pestgebete Des Morsili" (or the Pestilence of Morsili) in his book *Kleinasiatische Forschungen* on pages 161–251.[29]

The translation of the relevant section of the Plague Prayer about the pestilence follows: "When they led back the prisoners, whom they caught, to the land of Haitti, then in the midst of the prisoners pestilence took place. They began to die more and more".[30] The prisoners mentioned were Egyptian prisoners of war from the fighting between the Hittite Empire and the Egyptian Empire, where their spheres of influence met in Syria.

When the pestilence or Hittite Epidemic began in 1322 BCE its impact was devastating. For a start the Hittite king Suppiluliuma I died from the infection in 1322 BCE followed by his son and heir a year later, when he was King Arnuwanda II. This then brought Suppiluliuma's younger son Mursili II to the throne. He was to reign until 1295 during which time he had to watch as the infection ravaged his empire for 20 years killing off so many people that eventually, as he states in one of his many Plague Prayers, there were not enough farmers to plough the fields resulting in famine, which only compounded things.

The latest collection of Hittite Prayer translations is by Itamar Singer who is Professor of Ancient Near Eastern History and Culture at Tel Aviv University's Department of Archaeology. These will be used as the definitive translation as they have taken into account all previous attempts at translation by the Hittite scholars such as Forrer, Goetze and Guterback. Singer's qualifications are BA from Hebrew University, MA from Tel Aviv University and a PhD from Marburg University and Tel Aviv University. Singer's latest book, published in 2002, *Hittite Prayers* contains all of King Mursili's Plague Prayers.[31] Their review follows.

Hittite prayers to the gods were restricted to the kings mainly and in rare circumstances to other members of the Royal family. Prayers by the ordinary people such as seen in Babylon and Egypt, are not found in the Hittite Empire. The King prayed for himself, his health and for the health of his family but he mostly prayed for his land and his people. Mursili's Plague Prayers are a good example of how a King prayed for his people "Mursili's ultimate argument for divine mercy is the perishing of Hatti's population in the plague, rather than his own safety".[32]

Of all the Hittite prayers found so far there are none by the Great King Suppiluliuma I, but his son Mursili II is the most prolific of the Hittite kings. Mursili II wrote at least twelve prayers, two-thirds of which dealt with the plague and his enemies, and one-third dealt with the death of his wife Gassuliyawiya due to an illness (? the plague) in about the ninth year of his reign.

Did Mursili actually compose these prayers or did he leave it to some priest or ancient speech writer? Hans Gustav Guterbock (1908–2000) was an archaeologist and Hittitologist at the University of Chicago's Oriental Institute. Eventually he became the Tiffany and Margaret Blake Distinguished Service Professor of Hittitology. In 1976, he coedited the *Chicago Hittite Dictionary*. Guterbock thought Mursili was very involved in the composition of his prayers: "Mursili's personality speaks so vividly from the texts that go under his name that it is obvious that he must at least have given directions for their formulation, if he did not actually dictate them".[33]

These prayers lay bare the soul of the king, as they cry out for divine help and intervention. As Singer states:

> In face of military catastrophe, grave illness, or the death of a loved one, man disregards the conventional rules of prudent phrasing, and cries out from the bottom of his heart for deliverance and for a better comprehension of his world. We discover in these prayers many a moralistic reflection and poetical gem, often characterized by a deeply pessimistic concept of life.[34]

The prayers have a set format. The Defendant is either the King or His representative. The Prosecution is to the Offended God or Gods. To the Hittites:

> Every human sin is an offence against the divine world, but apparently there is a specific god who carries the complaint to the assembly of gods. The identity of this angry god of heaven or earth is usually not known to the supplicant, and the all-knowing Sun-god is invoked to find him and to intercede on the supplicant's behalf.[35]

The defense is the actual plea to the gods. "In principle, all the gods addressed in prayers may be considered as intercessors transmitting the king's plea to the assembly of gods convened in Hattusa".[36] Hattusa being the Hittite Empire's capital also had the main temple where the Gods assembled.

Thus each prayer has first an Invocation in which a specific god such as the all-knowing Sun Goddess is summoned to hear the prayer. Second there is the Hymn section when the summoned god is praised and flattered to get them in a good mood to finally hear the third section, the Pleading. In the pleading section the King states his case of what is wrong and what he is asking from the gods. The pleading by Mursili in his Plague Prayers is very touching and heart wrenching. He is begging the gods to stop this plague that has lasted 20 years and killed so many of his people including his father, his brother and maybe even his wife. Mursili, in the prayers, blames the sins of his father for the wrath of the Gods in the form of the plague.

The seven prayers by Mursili concerning Plague and Enemies, including the Plague Prayers, will be presented and numbered as they appear in Singer's book *Hittite Prayers*. The CTH number means the Catalogue Des Textes Hittites as devised by the French Hittitologist and Director of the French Archaeological Institute in Istanbul (later to become the French Institute of Anatolian Studies), Emmanuel Laroche in 1971.[37]

Plague Prayers 8–14 are shown completely at the back of this book.

No.8 Mursili's Hymn and Prayer to the Sun-Goddess of Arinna (CTH 376.A)

In this prayer Mursili complains not only about the terrible effects of the plague but also about revolt in neighboring states, such as the newly subjugated protectorates of Mittanni and Arzawa, and smaller vassal states such as the northern state of Kaska and Arawanna, Kalasma, Lukka and Pitassa to the west. In the pleading section he says that the plague is killing the people of Hatti so they are unable to make offerings of bread or libation to the gods. In other words the plague is being counterproductive resulting in the gods not being fed.

> O Gods What is this that you have done? You have allowed a plague into Hatti, and the whole of Hatti is dying. No one prepares for you the offering bread and the libation anymore. The plowmen who used to work the fallow fields of the gods have died, so they do not work or reap the fields of the gods. The grinding women who used to make the offering bread for the gods have died, so they do not [make] the god's offering bread any longer".[38] This is followed by a cry of pity "O gods [again] have pity on the land of Hatti. On the one hand it is oppressed with the plague, [and on the other] it is oppressed by hostility.[39]

No.9 Mursili's Hymn and Prayer to the God Telipinu (CTH 377)

This prayer is similar to number 8 but shorter. In the pleading, he asks the god Telipinu to keep him alive as well as his queen and princes, presumably because of the threat of death from the plague "keep alive the king, the queen and princes, and give them life forever, health, longevity and vigor!".[40]

Mursili then refers to the plague as the "evil fever"[41] and mentions famine due to so many people dying that no one can tend the fields. "But from Hatti [drive out] the evil fever, plague, famine and locusts".[42]

No.10 Mursili's "Third" Plague Prayer to the Sun-goddess of Arinna (CTH 378.111)

This is a short prayer because there is no Hymn section and only a small invocation section because a lot of writing was lost due to damage to the cuneiform tablet. In this prayer Mursili again states that the Gods "allowed a plague into Hatti" and "people kept dying",[43] but he also says that the Plague had been going on for 20 years: "for twenty years now people have been dying (in great numbers) in Hatti".[44]

Mursili also tells how it kills children and adults "[if someone] produces a child, [the] of the plague [snatches (?)] it from him. Should he reach adulthood, he will not attain old age ...".[45]

No.11 Mursili's "Second" Plague Prayer to the Storm-god of Hatti (CTH 378.11)

This prayer to the Storm-god of Hatti is Mursili's largest and best known Plague Prayer. In this prayer he says that Egyptian prisoners of war and civilian captives brought back to Hatti by his father Suppiluliumas I after fighting in the land of Amqa had introduced the plague.

> But when the prisoners of war who had been captured were led back to Hatti, a plague broke out among the prisoners of war, and [they began] to die. When the prisoners of war were carried off to Hatti the prisoners of war brought the plague into Hatti. From that day on people have been dying in Hatti"[46]

> "for this plague [caused by (?)] the prisoners of war who were brought back from Egyptian territory and by the civilian captives who were brought back.[47]

In this prayer Mursili also refers to what he thinks is the cause of the plague—three sins his father had made that would have offended the gods. Mursili then saw his job was to ask for forgiveness from the gods and to "keep making restitution for them [the three serious sins]"[48]

The first sin was that Suppiluliumas I had broken a treaty or oath made by the Storm-god of Hatti, between the men of Hatti and the Egyptians, when he invaded the land of Amqa. The subsequent fighting produced the plague infested Egyptian prisoners of war who started the plague.

The second sin was that Suppiluliumas I did not perform the Ritual of the Mala River that Mursili subsequently performed to appease the gods. This ritual is where offerings were made to the gods at the Mala River in a lavish ceremony.

The final sin was the murder of Tudhaliyas III, also known as Tudhaliya the younger, in 1344 BCE at Suppiluliumas II's command so he could become King himself. This is another example of how the extended Hittite Royal family fought amongst themselves.

These were all serious sins (breaking a promise with a god, offending the gods and murder) worthy of a 20-year plague as punishment according to Mursili.

No.12 Mursili's "First" Plague Prayer to the Assembly of Gods and Goddesses (CTH 378.1)

In this prayer "Mursili personally addresses the entire pantheon, divided into male and female deities, without listing their names as in later pleadings before the divine assembly".[49]

Mursili refers to the murder of Tudhaliya "[But when my father wronged Tudhaliya]"[50] and what affect the plague has had on his Empire "now the plague [has become] even [worse]. Hatti has been [severely] damaged by the plague and it as been decimated".[51] So the plague must have been having a seriously adverse effect on the Hittite Empire even to the point of the Empire becoming smaller. "Hatti has been oppressed by the plague, it has been reduced in size".[52] Mursili also laments how the plague continues on with no sign of ending "[The plague] does not subside at all, and they continue to die".[53]

He then goes on to say that even the people making offerings to the gods are dying "they continue to die, [even those] few [makers of offering bread] and libation pourers [who still remain will die, and nobody will prepare] for you offering bread and libation any longer"[54] and ask for pity "let me elicit your pity".[55]

Finally, Mursili asks the gods to send the plague to his enemies "send the plague [away from Hatti] ... O gods, my lords, turn the plague [away, and send] whatever is evil to the enemy land".[56]

No.13 Mursili's "Fourth" Plague Prayer to the Assembly of Gods (Arranged by Localities) (CTLH 378.IV)

This prayer is addressed to the full Assembly of Gods starting with the Noble Storm-god. The full list includes another thirty gods including especially the god of war and pestilence Iyarri (a god not be to missed under the circumstances)

In this prayer Mursili is trying to gain the favor of the gods by saying he will restore any neglected or destroyed temples in the Empire. He is also becoming increasingly desperate saying "O gods, my lords, you have turned your back on mankind".[57] He also keeps talking about mankind in general and not just the people of Hatti; so presumably the plague was killing people outside the Hittite Empire as well. "Mankind was [reduced in number] by plague, and your [servants] were reduced in number".[58]

Later Mursili reinforces his father's war with Egypt saying "When my father went to Egyptian territory, since that day of Egypt, death has persisted in [Hatti], and from that time Hatti has been dying".[59]

In the end he again asks the gods to send the plague away and let it subside in Hatti "Send the plague away from the land! Let it subside in the towns where people are dying, and let the plague not return to the towns in which it has subsided!".[60]

No.14 Mursili's "Fifth" Plague Prayer to the assembly of Gods (Arranged Typologically) (CTH 379)

In this prayer "Mursili makes a concerted effort to absolve himself of any misdeed that might have led to the plague".[61] According to Mursili it was still due to his father's three serious sins as he was only a child at the time. This is a short prayer because most of the text on the cuneiform tablet is missing.

In summary, the Plague Prayers of Mursili tell us that some sort of plague (of unknown cause) lasting 20 years ravaged the Hittite Empire, starting from 1322 BCE, when some infected Egyptian prisoners of war and civilian captives were transported from the land of Amqa into the land of the Hatti. The Plague Prayers also tell us that the plague was due to the wrath of the gods for the three serious sins of Mursili's father, and that Mursili was trying to appease the gods by carrying out the Ritual of the Mala River and by restoring any neglected or damaged temples.

It should also be noted that this plague not only killed Mursili's father, but it may have also killed his wife Gassulawiya in the ninth year of his reign (c.1312 BCE) and his son Piyassili in about 1315 BCE; as both suffered from fatal illnesses at the end of their lives and not from being murdered, accident or warfare.

How do we know that what has been translated from Hittite into English as the word "plague" actually means an infectious disease? First it was introduced by sick Egyptian POWs, second it lasted over 20 years and

Image 4.4 Photograph Hittite Carvings at Hattusa (© Magdalena Jankowska/iStock/Thinkstock)

third it was associated with a fever. All these factors point to an infectious disease of some kind.

So what could have infected the Egyptian prisoners of war who introduced the epidemic? As shown in the previous chapter—bubonic plague.

Notes

1. Gurney, D.R. *The Hittites*, (London: The Folio Society, 2005) 205–206.
2. Bryce, T. (op cit., 2010) 331 + (29).
3. Ibid +(30).
4. Ibid, 300+ (31).
5. Ibid, 343.
6. Ibid. Singer (77).
7. Ibid. + (81).
8. Ibid, 345+Seeher (82).
9. Ibid, 344.
10. Hare, R., *Pomp and Pestilence*, (Publisher Gollancz, 1954) 74.
11. Hare, Ronald, Archives in London and the M25 area. Identity Statement. Reference Code GB0120PP/HAR. (1899–1986).
12. Hare, R., (1983) "*The Scientific Activities of Alexander Fleming, Other than the Discovery of Penicillin*", (Medical History, Vol 27, 1983) 347–372.
13. Hare, R., (op.cit, 1954) 210.
14. Annotations, *Plagues and the Fortunes of War*, (Lancet, Jan.8 1949), 68.
15. Fitzpatrick Trust, Royal College of Physicians at http://old.rcplondon.ac.uk//history-heritage/College-history-new/ Historical-lectures, 1.
16. Email from Mr. Martin Killen to Dr. Philip Norrie on Wednesday 21 December 2011 @ 2.23 am.
17. Obituary, W.H. Wynn, MD, MSc, FRCP, (British Medical Journal, June 23, 1956), 1489.
18. Cooray, MPM, *Epidemics in the Course of History*, (Ceylon Medical Journal, June-Sept.1965), 89.
19. Ibid.
20. Tyldesley, J., *Ramesses: Egypt's Greatest Pharaoh*, (Penguin Books, 2000) 73.
21. Hopkins, D., *The Greatest Killer, Smallpox in History*, (Chicago: University of Chicago Press, 2002).
22. Ibid, 16.
23. Ibid.
24. Hopkins (op.cit, 2002) 16.
25. http://www.cartercenter.org/news/experts/donald_hopkins.html,p.1.

26. Forrer, E.O., *The Hittites in Palestine*, (II Palestine Exploration Quarterly, Publisher Palestine Exploration Fund, Vol.69, 1937) 100–115.
27. Ibid, 104.
28. Ibid, 104–105.
29. Ibid, 105.
30. Ibid, 106.
31. Singer, I., (2002), *Hittite Prayers*, (Georgia, Atlanta: Society of Biblical Literature, 2002).
32. Ibid, 7.
33. Ibid, 14.
34. Ibid, 17.
35. Ibid, 7.
36. Ibid, 8.
37. Laroche, E., (19710 Catalogues des textes Hittites (Etudes et commentaries, 75), Editions Klinchsieck, Paris.
38. Singer, I., (op.cit, 2002), 52.
39. Ibid, 52.
40. Ibid, 55.
41. Ibid, 55.
42. Ibid, 55.
43. Ibid, 56.
44. Ibid, 56.
45. Ibid, 56.
46. Ibid, 58.
47. Ibid, 60.
48. Ibid, 59.
49. Ibid, 61.
50. Ibid, 61.
51. Ibid, 62.
52. Ibid, 63.
53. Ibid, 63.
54. Ibid, 63.
55. Ibid, 63.
56. Ibid, 63.
57. Ibid, 65.
58. Ibid, 65.
59. Ibid, 65.
60. Ibid, 66.
61. Ibid, 66.

CHAPTER 5

How Disease Affected the End of the Bronze Age

The Ancient Near East 1700–1150 BCE: A Geopolitical Background

The Ancient Near East was the region home to early civilizations that were so important in the development of human society, and is referred to as the "Cradle of Civilization". The Ancient Near East region covered the modern Middle East including Mesopotamia (modern Iraq), Ancient Egypt, Ancient Iran, Anatolia/Asia Minor (modern Turkey), The Levant (modern Syria, Lebanon, Israel, Palestine and Jordan) and the Arabian Peninsula. For the purposes of this book, the region has been expanded to include the Eastern Mediterranean region adjacent to the Near East which includes Greece, Crete and Cyprus.

The Ancient Near East was characterized by being the first to practice intensive year-round agriculture, developing the first writing system, inventing the potter's wheel and mill wheel, pioneering centralized governments, law codes and empires. The Ancient Near East also introduced organized warfare, slavery and the stratification of societies, as well as laying down the foundations of the fields of astronomy and mathematics.

The Ancient Near East and Middle East Periodization Chart (Table 5.1) shows Near Eastern and Middle Eastern history divided into eras.

The focus of this book is the period from 1700 BCE when the Hittite Empire began until 1150 BCE when the Bronze Age in the Near East officially ended (the end being between 1200 to 1150 BCE). As can be seen

P. Norrie, *A History of Disease in Ancient Times*,
DOI 10.1007/978-3-319-28937-3_5

Table 5.1 Ancient Near East and Middle East Periodization Chart

Copper Age	Chalcolithic (4500–3300 BCE)	Early Chalcolithic	4500–4000 BCE	Ubaid period
		Late Chalcolithic	4000–3300 BCE	Ghassulian, Uruk period, Gerzeh, Predynastic Egypt
Bronze Age (3300–1200 BCE)	Early Bronze Age (3300–2000 BCE)	Early Bronze Age 1	3300–3000 BCE	Protodynastic to Early Dynastic Period of Egypt
		Early Bronze Age II	3000–2700 BCE	Early Dynastic Period of Sumer
		Early Bronze Age III	2700–2200 BCE	Old Kingdom of Egypt, Akkadian Empire
		Early Bronze Age IV	2200–2100 BCE	First Immediate Period of Egypt
	Middle Bronze Age (2000–1550 BCE)	Middle Bronze Age I	2100–2000 BCE	Middle Kingdom of Egypt and end of Sumer
		Middle Bronze Age IIA	2000–1750 BCE	Minoan civilization and Begining of End of Indus Valley Civilization
		Middle Bronze Age IIB	1750–1650 BCE	Second Intermediate Period of Egypt
		Middle Bronze Age IIC	1650–1550 BCE	Hittite Old Kingdom, Minoan eruption
	Late Bronze Age (1550–1200 BCE)	Late Bronze I	1550–1400 BCE	Hittite Middle Kingdom
		Late Bronze Age IIA	1400–1300 BCE	Hittite New Kingdom, Mitanni, Ugarit and End of Indus Valley Civilization
		Late Bronze Age IIB	1300–1200 BCE	(Dark Age, Sea Peoples)
Iron Age (1200–539 BCE)	Iron Age I (1200–1000 BCE)	Iron Age IA	1200–1150 BCE	Troy V II, Hekla 3 eruption
		Iron Age IB	1150–1000 BCE	Neo-Hittite states
	Iron Age II (1000–539 BCE)	Iron Age IIA	1000–900 BCE	Neo-Assyrian Empire
		Iron Age IIB	900–700 BCE	Kingdom of Israel, Urartu, Phrygia
		Iron Age IIC	700–539 BCE	Neo-Babylonian Empire

from this chart, this time period includes the Second Intermediate Period of Egypt, the Hittite Old Kingdom, Minoan Eruption, the Hittite Middle Kingdom, The Hittite New Kingdom, Mitanni and Ugarit Kingdoms, the Sea Peoples, Troy VII and the Hekla 3 eruption followed by the post End of the Bronze Age Neo-Hittite states in the remains of what was the Hittite Empire.

In 1700 BCE the Ancient Near East was dominated by Egypt in the Nile Valley and into the Levant, the Kingdom of Babylon in Mesopotamia and the Minoan civilization in Crete. Anatolia was made up of small city-states including Arzawa.

Egypt and Babylon's military tactics were based on foot soldiers and were no match for the hordes of invaders who came from the Central Asian plains and swept through the Near East. They came in fast two-wheeled horse-drawn chariots with a driver and an archer on board. This system of weaponry was developed and refined earlier, during years of warfare on the plains of Central Asia. This type of weapon or style of warfare was unknown among the classical oriental civilizations. The foot soldiers of Egypt and Babylonia were unable to defend against these invaders. In 1630 BCE, the Hyksos swept into the Nile Delta region, and in 1595 BCE the Hittites swept into Mesopotamia.

The Middle Bronze Age civilizations c.1700 BCE displayed all their characteristic social traits: a low level of urbanization due to an agricultural based society; small cities centered around royal palaces; numerous temples; a strict separation of classes between an illiterate mass of peasants and craftsmen and a powerful military elite. Knowledge of writing and education was reserved to a small minority of scribes and the aristocrats.

Developments in the ancient Near East after 1700 BCE include the Hittite Empire beginning c.1700 BCE, in c.1600 BCE the Minoan civilization reached its peak and Mycenae in Greece was occupied with the Mycenaean civilization flourishing until c.1200 BCE. In c.1450–1500 BCE the Mitanni Kingdom in northern Mesopotamia began, Minoan culture ended on Crete and the Kassites ruled Babylonia. Later the Mitanni Kingdom was divided in two with the western half taken over by the Hittite Empire and the eastern half taken over by the Assyrians.

At the end of the Bronze Age between 1200 and 1150 BCE only the Egyptian Empire remained with all other empires, kingdoms and city-states having been destroyed in the "catastrophe".

The Bronze Age and Iron Age: A Summary

Human cultural development can be measured by means of the types of materials used to make tools, utensils and weapons. First there was the Stone Age where hard flint stone was sharpened by chipping away and honing the cutting edge on another flint stone. This resulted in a sharp cutting edge, which could be used as a knife, axe or spear tip.

The next stage was the Copper Age, the first of the metal ages. Copper by itself is a relatively soft metal; so copper-only metal implements were weak. So the metal workers looked for ways to make the copper stronger. Through trial and error they arrived at the idea of mixing or alloying small quantities of tin with the copper to form bronze.

The dates for the Bronze Age varied from region to region depending on the region's copper and tin ore supplies or that region's ability to trade for bronze ingots. For example, the Bronze Age in the Near East (the region of interest in this book) went from c.3300–1200 BCE, while in South Asia it was from 3000–1200 BCE, in Europe 2300–600 BCE, in China 2000–700 BCE and in Korea 800–400 BCE. By contrast the Sub-Saharan African region bypassed the Bronze Age altogether going from the Stone Age directly to the Iron Age with the use of imported iron, because the region had no copper or tin mines and no ability to trade for bronze. Some cultures went straight from the Stone Age to the Iron Age, bypassing the Copper and Bronze Ages, when their culture was infiltrated by immigrants who brought Iron Age tools with them; such as when the British colonized Australia, New Zealand and Fiji.

Focusing on Ancient Near East, the Near East had the earliest Bronze Age because "Metallurgy developed first in Anatolia, modern Turkey. The mountains in the Anatolian highland possessed rich deposits of copper and tin".[1] Copper was also found in Cyprus, the Negev Desert and Iran. Tin was found in more distant places such as far off Cornwall (UK) and the North West Indian subcontinent, so trade in tin became a vital trade route commodity.

The Near Eastern Bronze Age has been divided into the Early Bronze Age (c. 3300–2000 BCE), Middle Bronze Age (c. 2000–1600 BCE) and the Late Bronze Age (c. 1600–1200 BCE) by scholars. A more detailed dating system is shown in Table 5.2.

The next stage in human cultural development was the Iron Age; this is the third and final stage in the usual three-stage sequence of Stone Age,

Table 5.2 Bronze Age dating

Bronze Age (3300–1200 BCE)	Early Bronze Age (3300–2200 BCE)	Early Bronze Age I	3300–3000 BCE
		Early Bronze Age II	3000–2700 BCE
		Early Bronze Age III	2700–2200 BCE
	Middle Bronze Age (2200–1500 BCE)	Middle Bronze Age I	2200–2000 BCE
		Middle Bronze Age IIA	2000–1750 BCE
		Middle Bronze Age IIB	1750–1650 BCE
		Middle Bronze Age IIC	1650–1550 BCE
	Late Bronze Age (1550–1200 BCE)	Late Bronze Age I	1500–1400 BCE
		Late Bronze Age IIA	1400–1300 BCE
		Late Bronze Age IIB	1300–1200 BCE

Pernicka, E. et al., "*Early Bronze Age Metallurgy in Northeast Aegean*", in Wagner, G.A. et al.(eds), *Troia and Troad: scientific approaches, natural science in archaeology*, (Berlin, London: Springer, 2003) 143–172

Bronze Age and Iron Age or the fourth in the more detailed five-stage sequence of Stone Age, Copper Age, Bronze Age, Iron Age and Steel Age. The Iron Age ended in each region with the beginning of what is known as the historical period, that is, the local production of sufficient detailed written sources and records.

Again the dates for the Iron Age vary from region to region depending on that region's access to iron ore mines. Again Anatolia was blessed with abundant iron ore in its mountains. Consequently the Ancient Near East had the earliest Iron Age from 1300 to 600 BCE followed by India 1200–200 BCE and Europe 1200–400 BCE, China 600–200 BCE, Korea 400–50 BCE and Japan 100 BCE–500 CE.[2] Other historians give earlier dates for the Iron Age in Anatolia saying the earliest iron ware dates from about 2000 BCE[3] and forged weapons at the latest by 1500 BCE.[4]

According to the late eminent Oxford historian John Morris Roberts when discussing the Hittite Empire, he stated it "enjoyed a virtual monopoly of iron in Asia, this not only had great agricultural importance but, together with their mastery of fortification and the chariot, gave the Hittites a military superiority which was the scourge of Egypt and Mesopotamia".[5] Also "the use of iron weapons by the Hittites was believed to have been a major factor in the rapid rise of the Hittite Empire".[6] By 1200 BCE, iron was in common use around the Middle East but it would take a few more centuries before it supplanted bronze as the dominant metal even though it was lighter and stronger.

One theory for the collapse of the Bronze Age was a lack of tin either due to it having been mined out or because its trade routes had been disrupted due to raiders,[7] thus forcing metalworkers to look for an alternative metal. Hence iron became the next preferred metal. Another explanation for the end of the Bronze Age is simply that iron, once discovered, was found to be a better metal because it was much stronger.

A Hittite tomb in Anatolia housed a dagger that had a smelted iron blade dating from 2500 BCE, making it one of the earliest ever examples of smeltered iron.[8] The Late Bronze Age saw the slow continuous transition from bronze to iron throughout the Near East region as price and availability of iron, not just the iron-making technology above, determined the rate of expansion of iron usage. Thus it was a gradual transition from late Bronze Age to early Iron Age over several centuries in the Near East around 1200 BCE.

Currently Accepted Causes for the End of the Bronze Age: A Summary

Introduction

The end of the Hittite Empire occurred c.1200 BCE[9] and was part of a much larger event—the end of the Bronze Age. The end of the Bronze Age was known as the "catastrophe"[10] because it signaled the end of all the known empires in the Eastern Mediterranean and Near East including the Hittite Empire, the Mycenaean Kingdoms and the Egyptian Empire in Syria and Canaan, but not the central Nile based Egyptian Empire. It was followed by the "Ancient Dark Age" when all that was glorious about these lost empires including, not only their palaces and cities but also their history, religions, writings, art, music, administration and organization, temples, libraries, political systems and trade routes were lost. This loss lasted many centuries until the emergence of the classical Greek, Assyrian and Persian Empires and the Hebrews later.

The catastrophe lasted from c.1200–c.1150 BCE and was characterized by its short time frame of about 50 years, mass migrations of populations and mass destruction where whole cities were destroyed and burnt, but also curiously many whole cities were just abandoned intact. It was also characterized by another mass population migration involving raiders called the "Sea Peoples" who invaded the Near East from the Mediterranean region. It was "violent, sudden and culturally disruptive".[11]

The Dartmouth University Archaeology Department has an online lesson about the collapse of the Mycenaean Palatial Civilization in Greece around 1200 BCE. It summarizes the causes of this collapse as: economic factors, climate change, internal social upheaval, invasion from outside the Aegean World and changes in the nature of warfare.[12]

Historian Robert Drews in his book *The End of the Bronze Age* has on his list of possible causes of the collapse the following: earthquakes, mass migrations, ironworking, drought, systems collapse, raiders and changes in warfare.[13]

The internet website, www.enotes.com, lists the possible causes for the Bronze Age Collapse, which includes the following: volcanoes, earthquakes, migrations, raids, ironworking, drought, changes in warfare and general systems collapse.[14]

Note that there is no mention of disease as a possible cause of the End of the Bronze Age in any of these three lists. By contrast this book proposes the more comprehensive list of sixteen causes for the end of the Bronze Age, including diseases.

But the final answer for the causes of the end of the Bronze Age will be multifactorial as some things will apply to one area, such as earthquakes in Greece that do not apply to other areas. A possible major cause though, one that knows no boundaries or political borders when it spreads are infectious disease epidemics. So what is the evidence?

Volcanoes

Mt. Hekla is a volcano located on the southern side of Iceland. According to volcanologists and Egyptologists, it erupted in 1159 BCE throwing an estimated "7.3 cubic km of volcanic rock into the atmosphere, placing its Volcanic Explosivity Index (VEI) at 5. This would have blocked out the sun leading to cooler temperatures in the northern parts of the globe for a few years afterwards. Thus plants would not grow causing famine and eventual disease in the weakened population. Traces of this eruption have been identified in Scottish peat bogs, and in Ireland a study of tree rings dating from this period has shown negligible tree ring growth for a decade".[15]

The Volcanic Explosivity Index is a relative measure of the explosiveness of a volcanic eruption on a scale of 1–8, with 8 being a mega-colossal eruption where the cloud column of ejected material would climb to a height of over 50 km. Such a large volcanic eruption could have caused famine in the Near East contributing to the demise of the region. The scale is logarithmic where each interval on the scale represents a ten-fold increase in the material ejected. By comparison the largest volcanic eruption in recent history was Mt. Krakatoa in 1883 which had a Volcanic Explosivity Index of 6, and Mt. Vesuvius which destroyed Pompeii in 79 had a Volcanic Exposivity Index of 5.

Comets

Mike Baillie is the Emeritus Professor of Paleoecology at the Queen's University in Belfast, Northern Ireland and is an expert of Dendrochronology, which is the science of dating by means of tree rings. He has built a year by year chronology of tree ring growth reaching back "the last 5,000 years".[16]

His work has confirmed "that the events such as those at 1628 BC, 1159 BC and AD 540, represented abrupt environmental downturns with profound effects on human populations".[17] These dates correspond with dynastic changes in China, the end of the Bronze Age in the Near East and the Plague of Justinian, respectively. These events were most likely volcanic in cause but he also argues that it could also be due to the impact of debris from comets blocking out the sun and causing temperatures to fall for years, hence poor growth in trees, as shown in their tree rings, and all other plants leading to famine which makes the general population more prone to disease. This has been formalized in his paper "Hints that Cometary Debris Played some Role in Several Tree Ring Dated Environmental Downturns in the Bronze Age".[18]

Earthquakes

Amos Nur is the Emeritus Professor of Geophysics at Stanford University, California, USA and he:

> postulates that earthquakes tend to occur in "sequences" or "storms" where a major earthquake above 6.5 on the Richter magnitude scale can in later months or years set off second or subsequent earthquakes along the weakened fault line. He shows that when a map of earthquake occurrence is

superimposed on a map of the sites destroyed in the late Bronze Age, there is a very close correspondence.[19]

In his paper "The End of the Bronze Age by Large Earthquakes?"[20] Nur proposes that a large earthquake storm lasting 50 years from c.1225 to c.1175 BCE could have contributed to the End of the Bronze Age.

Twentieth-century geophysical data about the geography of active tectonic faults especially at geological plate boundaries, the location of earthquakes, the geography of ground motion intensity, and earthquake frequency-magnitude statistics in the eastern Mediterranean show that most of the sites that collapsed at the end of the Bronze Age must have experienced destructive earthquakes repeatedly in their past. Recent and historical evidence shows also that these massive earthquakes reoccur every few hundred years in bursts, or "storms" of large events that sweep across broad portions (500–2000 km in length) of the eastern Mediterranean and over short periods of time (50 years).

This suggests that a large earthquake could have contributed to (and probably did contribute to) both the physical and political collapse of the great centers of civilization at the end of the Bronze Age. This probably began by an earthquake storm that unzipped the plate boundaries in the eastern Mediterranean between 1225 and 1175 BCE. The earthquakes in this 50-year long storm could have rendered many of the urban centers militarily vulnerable, thus inviting attacks, not by powerful distant Sea People but by opportunistic indigenous or neighboring populations. These attacks may have led in turn to the political and social collapse of the centres, followed by a dark age of recovery and rebuilding often lasting a few hundred years.[21]

Figure 8 of Nur's paper shows "Superposition of the sites destroyed at the end of the Bronze Age in 1200 BC and the earthquake intensity VII in our century (1910–1980). The coincidence suggests that the end of the end of the Bronze Age could be related to earthquakes".[22] (See the table of Bronze Age Earthquake Destruction in Nur's paper.)[23]

Drews in his book *End of the Bronze Age* dismisses the idea of earthquakes totally. He says that archaeologists only look at their own site and an earthquake may be relevant only to this site and not the whole of the Eastern Mediterranean. Nur, on the other hand, has shown that earthquakes occurred all over the Eastern Mediterranean and Near East.

Elizabeth B. French is an authority in Mycenaean Pottery and a former Warden of Ashbourne Hall, Manchester University and the former Director of the British School of Athens, from 1989 to 1994, which is a facility established in 1886 to promote British-based Greek archaeological research and study. In 1996 French wrote:

> Archaeologists of my generation, who attended university in the immediate aftermath of Schaeffer's great work in 1948, were brought up to view earthquakes, like religion as an explanation of archaeological phenomena to be avoided if at all possible. Thus, it is only recently that an earthquake at Mycenae has begun to be a serious hypothesis.[24]

Claude Frederic-Armand Schaeffer (1898–1982) was the first to conduct the excavations at the ancient city Port of Ugarit at Ras Shamra in Syria from 1929 that showed that the city was destroyed c.1190 BCE by the invading Sea People as told by Ammurapi, the King of Ugarit, in a letter to the King of Alasiya.

In 1948, Schaeffer first suggested that an earthquake may have destroyed late Bronze Age sites. His idea was rejected by archaeologists because the catastrophe was spread over a 50-year period and could not have been the result of a single earthquake. Schaeffer did not know then about 50-year long "earthquake storms" as proposed by Nur, which have put earthquakes back on the agenda. Schaeffer was a French scholar and archaeologist who was the curator of the Prehistoric and Gallo-Roman seum in Strasbourg (1924–1933) and the Museum of National Antiquities in Saint-Germain-en-Laye (1933–1956).

Fritz Schachermeyr (1895–1987) was an advocate for earthquakes playing a role in the end of the Bronze Age, which he proposed in his book *Griechische Fruhgeschichte*.[25] He has a similar view to Nur in advocating that catastrophic earthquakes destroyed Troy in Western Turkey, Mycenae in Greece and Knossos in Crete about 1200 BCE. Fritz was a professor specializing in ancient history and more specifically in Ancient Greek. He was a professor at Jena University (founded in 1558) in 1931, Heidelberg University (founded in 1386) in 1936,University of Graz (founded in 1585) in 1940 and finally the University of Vienna (founded in 1365) in 1952 until his retirement.

Tsunamis

Tsunamis usually accompany an underwater earthquake or volcanic eruption. So in line with the above evidence about earthquakes and volcanoes then a destructive tsunami may also have occurred, such as when the volcano on the island of Thera erupted sending a massive tsunami to Crete destroying many Minoan coastal settlements.

Mass Migrations of Populations

During the catastrophe many new ethnic groups started to appear in the Near East region.

Indo-European tribes such as the Phrygians, Thralians, Proto-Armenians, Macedonians and Dorian Greeks seem to have arrived at this time—possibly from the north. There also seems to have been widespread migration of the Aramaeans, possibly from the South East.[26]

The people involved in the mass migrations could have been "pushed" by drought and subsequent famine, earthquake destruction, raiders, and "Sea People" invasions. They could also have been "pulled" by the prospect of better lands and food, plus safety being far away from raiders. Another one of the possible "push" scenarios could be fleeing disease, and this should be factored in by historians.

Raids and Invasion by the "Sea Peoples"

The "Sea Peoples" is the name given to a confederacy of seafaring raiders who invaded the Eastern Mediterranean at the time of the catastrophe. They invaded Cyprus, the Hittite Empire and the Levant. They also tried to invade Egypt but were repelled. The Pharaoh Merneptah (reigned c.1213–c. 1203 BCE) referred to them as "the foreign countries or peoples of the sea" in his Great Karnak Inscription.[27] He fought and defeated them at Perire, in the western Nile Delta during the fifth and sixth years of his reign.

Later the Great Pharaoh Ramesses III (reigned c.1182–c. 1151 BCE) had to deal with a later wave of invasions by the Sea Peoples. The following inscription from his Medinet Habu mortuary temple describes what happened, who they conquered and who made up their confederation force:

> The foreign countries (i.e. Sea Peoples) made a conspiracy in their islands. All at once the lands were removed and scattered in the fray. No land could stand before their arms: from Hatti, Qode, Carchemish, Arzawa and Alashiya on, being cut off (i.e. destroyed) at one time. A camp was set up in Amurru. They desolated its people, and its land was like that which has never come into being. They were coming forward toward Egypt, while the flame was prepared before them. Their confederation was the Peleset, Tjeker, Shekelesh, Denyen and Weshesh, lands united. They laid their hands upon the land as far as the circuit of the earth, their hearts confident and trusting: 'Our plans will succeed![28]

Ammurapi was the last Bronze Age King of Ugarit. He wrote a letter to the king of Alasiya telling how the Sea Peoples had invaded and destroyed his country:

> My father, behold, the enemy's ships came (here); my cities (?) were burned, and they did evil things in my country. Does not my father know that all my troops and chariots(?) are in the Land of Hatti, and all my ships are in the Land of Lukka? ... Thus the country is abandoned to itself. May my father know it: the seven ships of the enemy that came here inflicted much damage upon us?[29]

The Sea People, like the other mass migrations, may have been "pushed" out of their lands by earthquakes, drought, and famine; and "pulled" to new lands looking for food and minerals such as iron, tin and copper. They too may have been "pushed" out of their lands by infectious disease. There are many hypotheses about who were the "Sea Peoples".

1. *Philistine hypothesis*: The Hebrews called the southern coastal plain of ancient Palestine "Philistia" and the people who lived there in the five towns called the "Philistine Pentapolis" of Askelan, Ashdod, Ekron, Gath and Gaza were called "Philistines". There are two schools of thought as to their origin. One school says they were Mycenaean because they made distinctive Mycenaean IIIC pottery, whereas the other school says they were an indigenous Canaanite culture. Either way they became a force that helped attack Egypt as part of the "Sea Peoples".
2. *Minoan hypothesis*: After the eruption of the volcano at Thera (today's Santorini), sometime between 1660–1613 BCE, Crete would have been devastated by fires and ash causing a cooling climate resulting in crop failures and famine, plus the tsunamis

destroying the coastline. The people of the Minoan culture would have been forced to leave Crete, some scholars say initially to Anatolia and from there later to the Levant as "Sea Peoples".

3. *Greek migration hypothesis*: Greece not only possibly supplied people for the Philistines but also supplied people who migrated to Sicily and Sardinia as well as fought in Troy and later occupied Cyprus. So they would have been bands of raiders/invaders like the Sea Peoples.
4. *Trojan hypothesis*: Historian Eberhard Zangger proposes that the "Sea Peoples" may have come from Troy and its allies, and that the Greek literary tradition of the Trojan War, as told by the poet Homer, may well reflect the Greek efforts to counter those raids.[30] Zangger (1958) is a Swiss German who has a PhD in natural sciences from Stanford University and was a senior research associate in the Department of Earth Sciences at the University of Cambridge from 1988–1991. He is a writer on geoarchaeology especially in the prehistoric Aegean area.
5. *Mycenaean warfare hypothesis*: In this hypothesis the Greek Mycenaean city states fought each other over decades. Refugees from this fighting then turned to piracy for survival and that this in turn led to becoming "Sea Peoples" raiders.
6. *Italian peoples hypothesis*: Italy may have supplied some of the Sea Peoples from the Etruscans.
7. *Anatolian famine hypothesis*: In this hypothesis extensive drought in Anatolia, which led to famine, in turn forced people from places in Anatolia such as Lydia to go to sea and become sea-going migrants. These displaced people could later become Sea Peoples.

   A famous passage from Herodotus portrays the wandering and migration of Lydians from Anatolia because of famine: In the days of Atys, the son of Manes, there was a great scarcity through the whole land of Lydia … So the King determined to divide the nation in half … the one to stay, the other to leave the land ... The emigrants should have his son Tyrrhenus for their leader ... They went down to Smyrna, and built themselves ships … After sailing past many countries they came to Umbria ... and called themselves … Tyrrhenians.[31]
8. *Invader hypothesis*: Historian Michael Grant proposed that there may have been another origin for the "Sea Peoples" besides the Aegean. He proposed that "There was a gigantic series of migratory waves, extending all the way from the Danube valley to the plains of China."[32] Grant (1914–2004) received a Doctor of

Literature from the University of Cambridge and was an English classicist, numismatist and author of more than seventy books on ancient history and ancient coins.

Fellow historian Sir Moses I. Finley has suggested that the Carpatho-Danubian region of Europe was the original centre of this disturbance. Sir Michael Grant and Sir Moses I. Finley suggest that these people were invaders who destroyed sophisticated cities and built simpler and less complex settlements with plain pottery and simple tools, on top of the ruins. This demonstrates a cultural discontinuity[33] from a more advanced culture to a less advanced culture after the invaders or raiders took over. Sir Moses I. Finley (1912–1986) was born in the USA as Moses Israel Finkelstein but changed his surname to Finley in 1936. He studied at Syracuse University and Columbia University and taught ancient history at Columbia University and Rutgers University in the USA, specializing in the social and economic aspects of the ancient world. After being attacked for his left wing opinions by Senator Joseph McCarthy in 1954 he immigrated to Britain where he taught classics at the University of Cambridge from 1955, eventually becoming professor of ancient history from 1970 to 1979 when he was also knighted.

Ironworking

The collapse of the Bronze Age should also be seen as part of the bigger technological picture and changes taking place at the time, that is, the slow change from bronze making to iron working. Even though the Hittites in Anatolia were the first great power to have iron at the time of the collapse, the general regional shift from bronze to iron occurred after the collapse of the Bronze Age c.1200 BCE.[34] The Sea Peoples' weapons for example were made of bronze, not iron. So iron confirmed the collapse and end of the Bronze Age but did not cause it.

Lack of Tin

Bronze is made from a combination of 85–95% copper and 5–15% tin. In the ancient world tin was an element in short supply. Without tin the only metal available for implements and weapons was the much softer plain copper of the Copper Age.

The Near East had few tin mines with Anatolia being the main source of tin. The majority of the tin was imported from outside the Near East, with Cornwall in Britain and other sources to the distant East of the Near East, such as Afghanistan, being the main suppliers. Hence the tin trade and its trade routes were very important, like oil trade today. If the tin trade collapsed for whatever reason, such as war or disease, then the Bronze Age would finish, as people would have to look for alternative metals such as iron, also the Bronze Age could have ended because iron is a stronger hence better metal and would have replaced bronze as a metal of choice.

Drought

Herodotus (c484–c425 BCE) was a Greek Historian who is regarded as the "Father of History". He wrote a detailed history of the Greco-Persian wars called the "Histories" and he "was the first historian known to collect his materials systematically, test their accuracy to a certain extent and arrange them in a well-constructed and vivid narrative".[35] He proposed that the Bronze Age ended because of a long drought that caused famine and social disruption.

Examples of grain shortages in the Hittite Empire and demands for urgent resupply of Hatti in the late Bronze Age, just before the end, were mentioned in the last chapter. Dendrochronological (tree ring) studies have shown drought years leading up to 1200 BCE. "Starting from 1209 BCE, we see indices smaller than 100 per cent, except for two years, 1205 and 1204 BCE. In addition there is a continuous decrease in indices in the last five years of the empire. Examining the tree rings shows the existence of dry years which contributed to drought on the lands".[36]

Peter Ian Kuniholm founded the Aegean Dendrochronology Project in 1973 and has been its Director ever since. This project examines tree rings all over the Aegean and Near East to make up a chronology of the area from Neolithic time to the present. He has shown that in 1190 BCE trees grew at only 53.5% of their normal rate, in 1189 BCE at 61.1% and in 1188 BCE at 62% indicating three consecutive years of drought, which would lead to famine. Examination of the data presented in this paper shows that the occurrence of three such poor growth years in a row was rare.[37] Kuniholm is the Professor of Archaeology and Dendrochronology at Cornell University. He gained his BA at Brown University in 1958, his

MA at Vanderbilt University in 1963 and his PhD from the University of Pennsylvania in 1977.

Other researchers also support the arid climate change theory at the end of the Bronze Age. Anthropologist Brandon Drake from the University of New Mexico studied various parameters including Mediterranean Sea temperatures to show the change in the climate to much drier times at the end of the Bronze Age and early Iron Age.[38] By contrast, archaeologist Jennifer Moody studied the change in building structure over the Late Bronze Age period to show that buildings had verandas, and were airy with good ventilation to keep them cool because the climate became "significantly drier than previously".[39]

Famine

Naturally what follows drought is famine. As discussed above, there was correspondence between the Hittite King and the Egyptian Pharaoh about urgent grain shipments for Hattusa. Famine can drive people from their lands and could be one of the causes for the mass migrations seen at this time, and could also have pushed the Sea Peoples from their land in search of new territories with food.

Famine can also weaken people and make them more prone to disease. For example prior to the Black Death, which started in Europe in 1347, there was the Great Famine which lasted from 1315–1317 which greatly weakened the population of Europe. The Great Famine was not caused by drought, as in the end of the Bronze Age though, but by several years of excessive rain fall. During this time crops were flooded or could not be planted. Disease could also cause famine because the farmers or slaves necessary to work the fields would have died, hence crops would not have been planted or maintained, let alone harvested, which required considerable man power.

Changes in Warfare

Robert Drews "suggests that these collapses occurred as the result of a fundamental change in the nature of warfare in this period".[40] The horse drawn chariot with an archer behind the driver was introduced in the seventeenth and eighteenth centuries BCE and subsequently dominated the Near Eastern battlefields. They did not dominate in the Aegean as the topography was different being more rugged terrain with fewer large flat areas suited to chariot warfare. These chariots were manned by a socially and economically privileged warrior elite (e.g. the Hittite Empire, the Mycenaean Kingdoms, city states of coastal Syria and the Levant (such as Ugarit), the Hurrian Kingdom of Mitanni, the Kassite Kingdom of Babylon etc).[41]

Drews argues that these large massed chariot formations involving sometimes thousands of chariots became vulnerable when they were harassed and killed by highly mobile, lightly armed infantry who used new and better weapons with great success. These new weapons included (a) the Aspis—a highly maneuverable small round shield 2–3 feet in diameter as opposed to the Sakos or large immobile shield; (b) the Javelin which had elliptical heads instead of a barb which allowed easy retraction hence reuse. Massed javelins thrown on the run using massed swarming formations were very effective against massed chariots; (c) The new Nave Type II sword which was about 70 cm long which "optimized both for slashing as well as stabbing or thrusting"[42] and finally (d) "The infantryman's corslet, which protected the trunk, arose, along with greaves for protecting below the knees".[43]

This new warfare meant that an entire elite social order of charioteers became redundant and were replaced by infantry and cavalry. This meant significant disruptions of the aristocracy and consequently the royal families.

Unfortunately Drews only provides a single answer solution to a very complex multicausal problem, hence is guilty of oversimplification. Also his approach does not apply to the Aegean where chariots were less dominant due to the hilly topography. After the end of the Bronze Age Mycenaean pottery continued to depict war chariots as part of the decoration thus confirming that they had not died out. This assumes that the potters were portraying only contemporary and not historical scenes.

General Systems Collapse

The late Bronze Age Palatial Kingdoms all had fatal centralized, complex and top-heavy political structures, which made them vulnerable. This bureaucratic domination and top-heavy administration involved the use of a whole scribal class of record keepers and was highly specialized. It also involved the exploitation of the peasantry who funded the whole expensive top-heavy system with their taxes.

So when the system was stressed by famine, war or disease for example, it was unable to cope due to lack of funding to prop up "the system". Once the administration was affected then other things such as trade became affected. Once trade in all important commodities such as tin, grain, olive oil, wine and timber was stopped, the end was inevitable. A weakened administration could also not handle other crises such as peasant revolts, defection of mercenaries, overpopulation and invasion by Sea Peoples.

The General Systems Collapse theory was first proposed by Joseph Tainter in his book *The Collapse of Complex Societies*.[44] He argued that societies collapse when their investment in their social complexity becomes unsustainable. This book proposes that one of the causes for the General Systems Collapse c.1200 BCE was infectious disease epidemics. Tainter (1949–) is an American anthropologist and historian who gained his PhD in 1975 from North Western University. He is currently the Professor of Environment and Society at the Utah State University.

Cost of Maintaining an Empire

Vast sums of money were needed to run an empire especially one with an expensive and large bureaucracy. Besides their administrative expense, there was the cost of the Royal Family and the large armies necessary not only to defend the empire but also to help expand it. Once the revenue from taxes dried up the system, lacking funding, collapsed. So this is a consequence of the the reduced income caused by peasants migrating and dying due to famine, war or disease, but not a cause of the end of the Bronze Age.

Economic Factors

This theory put forward variously by Vermeule (1960),[45] Lakovides (1974)[46] and Betancourt (1976)[47] suggests that the piratical activity of the "Sea Peoples" would cripple trade routes, hence disrupt commerce so much as to bring down the late Bronze Age empires. The theory refers to a consequence of the "Sea Peoples" activity and not to why the "Sea Peoples" were active in the first place. So once again there is a theory which turns out to be a "consequence of" and not a "cause of" the end of the Bronze Age.

Internal Fighting Within Ruling Royal Families

The Hittite royal families provide many examples of internal fighting within a royal family. Murder, even of the incumbent king as a means to gaining control of the throne, was accepted practice. Fighting between different branches of the royal families leading to civil war would have badly destabilized any empire.

Disease

William H. Stiebing, Emeritus Professor of History at the University of New Orleans, in his book *Ancient Near Eastern History and Culture* he stated in chapter 8 "The End of the Bronze Age" that "Something disturbed the fragile economic, social, and political systems of Bronze Age states and set off a series of chain reactions ... [That] ... destroyed the eastern Mediterranean Bronze Age civilizations."[48] Stiebing includes the word "plagues" in his list of possible causes but does not elaborate any more on the subject.

Marc Van De Mieroop, Professor of History at Columbia University, says in his book *A History of the Ancient Near East* that "No single cause can explain this comprehensive change."[49] When trying to explain the end of the Bronze Age.

Oliver Dickinson from Durham University had a similar view when he stated in chapter 36 "The Collapse at the end of the Bronze Age" in his contribution to *The Oxford Handbook of The Bronze Age Aegean (c3000–1000 BC)* that "it is a waste of effort to try to isolate a single cause or prime mover for the Collapse."[50] This is because it was most probably the result of several causes.

This is what this book is arguing that the end of the Bronze Age in the Near East, referred to as either "The Catastrophe" or "The Collapse" [Dickinson preferred the term Collapse], was the result of a cascade of events beginning with volcanic eruptions and comets. These caused drought which in turn caused famine, which weakened the population so they (rich and poor alike, because disease does not discriminate) were prone to whatever infectious disease happened to be prevalent at the time. This is in agreement with Dickinson's "Natural disasters like earthquakes, localized droughts leading to famines, or epidemic diseases could have acted as catalysts for trouble or have exacerbated an already deteriorating situation".[51] Dickinson also showed that many sites were abandoned at the end of the Bronze Age not reoccupied for many centuries. This would be expected when an infectious disease epidemic is active. His abandoned city list included places such as Gla, Messenia and Krisa in the Aegean, besides the well documented abandonment of Hattusa, the Hittite capital in Anatolia.

Guy Middleton's PhD, "Theories of Mycenaean Collapse", favored the idea of infectious disease as a cause of the collapse, so it included a section on "Plagues and epidemics" but stated "The main problem with the plague hypothesis is the lack of positive evidence".[52] It also stated "Much about the plague hypothesis seems attractive in the way it can explain the changes in settlement pattern, material and social cultural and long-term decline in population levels".[53] The plague hypothesis was also good for "explaining the variability of collapse[54] because "palatial areas may have been more densely settled and thus more seriously affected, while possibly less populous peripheral areas may have been less affected".[55] Once the central palace was adversely affected: "This lack of strong central power and instability seems to fit with the kind of unstable and more mobile society suggested for palatial Greece."[56]

Emeritus Robert Arnott from Oxford University wrote chapter 1 "Disease and the Prehistory of the Aegean" in *Health in Antiquity* edited by Helen King, in which he listed some of the possible diseases from that time which included: thalassemia, malaria, tuberculosis, brucellosis, malnutrition, scurvy, anemia, measles, chickenpox, oral infection, pneumonia, hemolytic disease, enteropathies, cholera, typhoid, dysentery, tetanus, hookworm, mumps, whooping cough and amoebiosis. Note that most diseases listed are infectious diseases. He also gave a possible reason why disease has not been considered as a cause for the end of the Bronze Age because "many such scholars are completely unaware of the social effects of disease and the major consequences that ensued whenever contacts across disease boundaries allowed a new infection to invade a population that lacked any acquired immunity."[57]

Arnott referred to the research of J. Lawrence Angel (also known as the archaeological "bone man") and Robert Sallares who have found the lethal falciparum malaria DNA in ancient skeletons, making malaria a strong infectious disease that possibly contributed to the end of the Bronze Age. He also supported the drought leading onto disease hypothesis stating "If widespread climatic change and drought were mainly responsible for the decline and collapse of the Mycenaean world in the twelfth century BC, as has been suggested ... then a natural consequence would have been widespread epidemic disease (e.g. cholera and typhoid) brought on by the shortage of clean drinking water".[58] In conclusion, Arnott paints a very bleak picture of life for the ordinary man in ancient times "the harsh reality of a society where life was hard, death and disease were everyday occurrences and the day-to-day ambition of those who lived outside the palaces was simply survival".[59]

Eric Watson-Williams wrote an article about the end of the Bronze Age called "The End of an Epoch" in which he championed bubonic plague as the sole cause for the catastrophe. "What seems so puzzling is the reason why these apparently strong and prosperous kingdoms should disintegrate"[60] he questioned. He cites abandonment of cities, the adoption of the practice of cremation of the dead instead of the usual burial because so many people died, and it was necessary to destroy the decomposing bodies quickly, and the fact that bubonic plague is very deadly killing animals and birds as well as humans, affects large areas, spreads rapidly and lingers for many years as reasons for his choice of bubonic plague. Unlike Panagiotakopulu, he provides no physical evidence, but uses history to compare how things were during later bubonic plague epidemics such as the Plague of Justinian and the Black Death to how things were around 1200 BCE.

Watson-Williams argues that there were four epidemics of the bubonic plague that we know of in ancient times, namely the Hittite Epidemic of 1322 BCE, the Exodus from Egypt in the reign of Merneptah c.1230 BCE, the Plague of the Philistines c.1130 BCE and finally the plague "that which destroyed the army of Sennacherib (701 B.C.)".[61] He also quotes a pestilence that killed 70,000 men in three days during the reign of King David c.1017 BCE that may have been bubonic plague.

Lars Walloe from the University of Oslo had a similar view to that of Watson-Williams when he wrote his article "Was the disruption of the Mycenaean world caused by repeated epidemics of bubonic plague?" He noted the "large movements of population";[62] "The population decreased in successive steps during the first two or three epidemics of plague down to perhaps half or one-third of its pre-plague level",[63] and that there was "a substantial reduction in agricultural production".[64] This led to famine and the abandoning of settlements. He thus concluded that bubonic plague accounted for all of these observations rather than other infectious diseases such as anthrax.

Diseases that may have Caused the End of the Bronze Age and the End of the Hittite Empire

What Are the Most Likely Diseases That Could Have Helped End the Bronze Age and What is the Evidence?

Any historian trying to find the cause of the end of the Bronze Age and the Hittite Empire must explain: the short time frame of approximately 50 years, when it occurred between 1200–1150 BCE; the mass migrations not only of normal people but also of the "Sea Peoples"; and the fact that so many large cities, such as the Hittite capital Hattusa, were simply abandoned and not destroyed or occupied by raiders or invaders. Disease in the form of infection epidemics provides one plausible explanation; there is nothing like a severe widespread plague to deliver the final fatal blow to an empire or an era.

Bubonic Plague

The bubonic plague is caused by the *Yersinia pestis* and is transmitted to man by fleas from small rodents such as rats. It is a very lethal disease killing two out of every three people infected within four days. The usually accepted first outbreak of bubonic plague was the Plague of Justinian in 541. "The first recorded outbreak of bubonic plague was the world's first great pandemic. Called the Mortalitas Magna (Great Death) or the Plague of Justinian, it began in Arabia ... The pestilence reached Constantinople by the spring of A.D. 542 and lurked around the eastern Mediterranean until the 760s."[65]

New research however by Mark Achtman has suggested that the bubonic plague began near China c. 600BCE and spread to Europe via Central Asia's "Silk Road".

Mark Achtman's research was published in Nature Genetics in an article titled "*Yersinia pestis* genome sequencing identifies patterns of global phylogenetic diversity".[66] His team compared seventeen whole genomes of *Yersinia pestis* isolates from various global sources and "conducted phylogenetic analyses on this sequence variation dataset, assigned isolates to populations based on maximum parsimony and, from these results, made inferences regarding historical transmission routes. The phylogenetic analysis suggests that *Yersinia pestis* evolved in or near China and spread through multiple radiations to Europe, South America, Africa and Southeast Asia, leading to country-specific lineages".[67]

Mark Achtman BSc, MSc, PhD from the University of California, Berkeley, USA is a Canadian microbiology researcher at the University College, Cork, Ireland where he specializes in the population genetics of bacterial pathogens and microbial phylogeography. Prior to this he was a researcher at the Max-Planck Institute in Berlin, Germany specializing in molecular genetics and infections biology.

But, as shown earlier in this book, Kozloff and Panagiotakopulu argue that bubonic plague, coming from India [? Indus Valley Civilization] could have occurred during the reign of Amenhotep III early in the fourteenth century BCE, that is, predating Achtman's Chinese suggestion by over 750 years. Thus, if Achtman continued his research he may find that China may have contracted the plague from India, and that India was the primary source of the bubonic plague.

This all leads to the Plague of the Philistines, also known as the Plague of Ashdod, which occurred c.1190 BCE and may have been caused by the bubonic plague. If Egypt had bubonic plague during the reigns of Amenhotep III and his son, Akhenaten (c1370–1350 BCE) then it could have recurred to the north east in southern Canaan fifty years later.

The Philistines lived in south Canaan at the end of the Bronze Age and ruled their five city-states (Pentapolis of Gaza, Ashkelon, Ashdod, Ekron and Gath). They were the Kingdom of Israel's worst enemy and they fought each other many times. Their origin is obscure though. Some historians, such as Carl Ehrlich,[68] believe they were "Sea Peoples" from either the Aegean or Mycenae in Greece who settled in southern Canaan after being defeated by Pharaoh Ramesses III in c.1190 BCE, while others such as Riemschneider[69] believe they came from Anatolia as refugees from the crumbling Hittite Empire.

As stated earlier the Plague of the Philistines occurred c.1190 BCE[70] while others think it may have occurred later, either 1141 BCE or in the second half of the eleventh century BCE.[71] But Peter Kuniholm's Aegean Dendrochronology Project has shown that our current dating scheme for ancient times maybe anywhere up to 150 years out when compared to his tree ring dating scheme. "Long standing assumptions and conventions in other Egyptian and Old World chronology and history will need to be re-examined"[72] he has stated, hence the 1141 BCE stated above could in fact have been about 50 years out making it c.1190 BCE.

According to author Lee Allyn Davis in his book *Natural Disasters*, the Philistine Plague was caused by bubonic plague and began in 1200 BCE after the Philistines captured the Ark of the Covenant (a box that contained ancient Hebrew records such as the Ten Commandments and the First Torah Scroll) from the Israelites.

The first recorded plague is the one which beset the Philistines in 1200 BCE, and which is recorded in the Bible in the Book of Samuel. The Philistines in this year defeated an army of nomadic Hebrews at Ebenezer, captured the sacred Ark of the Covenant and carried it in triumph to Ashdod, a city near the Mediterranean Sea. But their triumph was immediately tainted, according to 1 Samuel 5:9: "the hand of the Lord was against the city with a very great destruction; and he smote the men of the city, both small and great, and they had Emerods (swellings) in their secret parts".

The description makes it clear that bubonic plague had invaded the army of Philistines, probably from a stricken ship. If it had originated in the Ark of the Covenant as the Bible notes, it would have been mentioned in the Old Testament.

Wherever they took the Ark (of the Covenant), the Philistines took plague too. They moved from Ashdod inland to Gath, then to Ekron. The plague followed them. Terrified, they trundled the Ark of the Covenant into a cart pulled by two milk cows. If the cows took the Ark to the Hebrew border town of Beth-shemesh, they reasoned that the Lord of Israel was responsible for the plague, and had indeed smitten them.

The cows took the Ark into the field of Beth-shemite, Joshua stopping alongside a huge stone. Israel rejoiced, but not for long. In 1 Samuel 6:19, the Bible chronicles the inexorable progress of the plague; "and he smote the men of Beth-shemesh, because they had looked at the Ark of the Lord, even He smote the people fifty thousand and three score and ten men; and the people lamented because the Lord had smitten many of the people with a great slaughter".[73]

Davis gets his idea of the plague coming from a stricken ship and not the Ark of the Covenant from a book about plague called *Plague an Ancient Disease in the Twentieth Century* by Charles T. Gregg.[74] In it Gregg states that the Plague of the Philistines occurred in the twelfth century BCE and that:

> The plague of the Philistines probably invaded the town of Ashdod from a stricken ship rather than with the Ark of the Covenant and then infected the crowds that conveyed the ark to the other afflicted cities. Had the plague begun in Israel there should have been accounts of it in the Old Testament, but Samuel mentions the disease no more.[75]

Others who also think the Plague of the Philistines was bubonic plague include W.J. Simpson and W.W.C. Topley and G.S. Wilson.

Simpson (1905)[76] affirms that the pestilence was bubonic plague, and that the "emerods" were plague buboes, and his assertion has been repeated by later writers. Topley and Wilson (1946),[77] for example, assert that "In the fifth and sixth chapters of the First Book of Samuel there is an unmistakable account of bubonic plague."[78] Simpson served as a health officer in Calcutta and Hong Kong, and was a member of the Royal Society of Tropical Medicine and Hygiene. He was also an MD from Aberdeen, a PRCP (London), DPH (Cambridge), Professor of Hygiene, Kings College, London and Lecturer in Tropical Hygiene at the London School of Tropical Medicine. In 1905 he published his book *A Treatise on Plague* through Cambridge University Press.

The description of the plague also mentions tumors that could have been buboes or the swollen lymph nodes associated with bubonic plague. Watson-Williams[79] and Walloe[80] also support the theory that bubonic plague caused the Hittite Epidemic of 1322 BCE, the Plague of the Philistines c.1190 BCE and the end of the Bronze Age c.1200 BCE. From its bases in Egypt and Canaan the bubonic plague could have spread across the entire Near East and Eastern Mediterranean region destroying empires and ending the Bronze Age.

Wilson MD, FRCP, DPH Director of Public Health Laboratory Service, England and Wales was a British bacteriologist, a member of the UK Whooping Cough Immunisation Committee in 1954, and wrote the Foreword to Hugh Paul's book *Control of Communicable Diseases* published in 1953. Topley (1886–1944) was a British bacteriologist who became Fellow of the Royal Society in 1930. Together Wilson and Topley wrote the standard textbook on the subject of Pathology namely *Topley and Wilson's Principles of Bacteriology and Immunity* in 1946.

Dysentery

In his book on the *Plague of the Philistines*[81] Shrewsbury goes against the conclusions of his fellow British bacteriologists Simpson, and Topley and Wilson and instead endorses the belief of the first century Romano-Jewish historian Titus Flavius Josephus (37–c.100 CE) who stated "categorically that it was dysentery, and he also discriminates the epidemic from the simultaneous plague of mice".[82] This idea was revealed by W. Whiston who translated the works of Flavius Josephus and had them published in 1793.[83]

John Findlay Drew Shrewsbury (1898–1971) was Professor of Bacteriology at the University of Birmingham from 1937 to 1963. He wrote on the historical aspects of infectious diseases especially about the plague writing *A History of the Bubonic Plague in the British Isles* reprinted in 2005 by Cambridge University Press, *Epidemic Disease and History* in 1947 and *The Plague of the Philistines, and other Medical-Historical Essays* in 1964 for example. Shrewsbury writes:

> It is possible that Josephus had access to authentic material that has since been lost; it is certain that he was many centuries nearer the ancient Jewish tradition than we are, and, though it appears to be fashionable in some quarters to decry the value of tradition as an adjunct to history, it is not wise to ignore it. With the support of Josephus, I submit, therefore that the plague of the Philistines was one of the forms of bacillary dysentery, either Shiga or a virulent Flexner dysentery, and that the "emerods" were "piles."[84]

Note his point that "it is not wise to ignore it". In other words he is warning us that it is not wise to ignore the ideas of Josephus.

The basis for his findings were the facts that he thought firstly the "secret part" of the body as the only part not visible when naked, that is, the anal area and that "hemerods" were in fact hemorrhoids or piles, which would be made worse with dysentery. This is in contrast to the others who thought the "secret part" was the genital area in general hence the "hemerods" were buboes or the swollen lymph nodes in the groin; because buboes are inflamed lymph nodes and there are no lymph nodes in the anal area.

The communicable diseases that have scourged armies throughout historical times are typhus fever, typhoid fever, and dysentery, and there can be no doubt that the pestilence that afflicted the Philistines first erupted in their army, and was then disseminated by their victorious troops among the civilian population of their cities. Typhus fever and typhoid fever can be dismissed, because neither disease is accompanied by the development of swellings around the anus; but bacillary dysentery is frequently associated, because of the invariable concomitant tenesmus, with the formation of external piles, those rounded protuberances from the anal margins that are at first hot, red and painful, but which later often itch considerably as they resolve. It is commonly stated in medical textbooks that bacillary dysentery in hot climates is more prone to give rise to piles than is the disease in temperate ones. The usual explanation given is that the anal sphincter is generally more lax in individuals living in hot lands than in those living in temperate regions.[85]

But Shrewsbury gives another reason for his conclusion that the Plague of the Philistines was in fact dysentery and not bubonic plague.

There is still the third clue to be considered; to wit, the fact that the Gethrites, after deliberate consultation, made "seats of skins" for themselves. To those who still opine that the pestilence of the Philistines was bubonic plague, the question may be put: Why should any sane individual make themselves seats of skins as a palliative for bubonic plague? Of what conceivable use would a skin seat be to a person who is collapsed, delirious, and bed-ridden almost from the onset of that disease? If the "emerods" were plague buboes, the action of the Gethrites was utterly nonsensical; but if, as I have argued, they were hemorrhoids, then anyone who has suffered from external piles will agree that the Gethrites showed good sense. The difference in comfort to the sufferer from piles between reclining upon a skin seat and sitting cross- legged upon the ground must be experienced to be appreciated, but I can assure the reader that it is profound.[86]

The significance of the plague cannot be underestimated. Because of this plague the Philistines returned the Ark of the Covenant back to the Israelites to get rid of it, which consequently allowed them to flourish.

Whatever opinion is held about the nature of the pestilence that ravaged the Philistines, the fact that it caused them to return the Ark of the Covenant to the defeated and demoralized Israelites is indisputable. That restoration deprived the Philistines of the inestimable moral advantage that the possession of the Ark gave to them, and at the same time restored to the Israelites the focus upon which all their national activities and aspirations converged. I therefore affirm that this act, because it saved the Israelite society from almost certain extinction, must be ranked as one of the decisive acts in human history, and I draw support for that affirmation from the authoritative pronouncement of Osterley and Robinson (1932): "Yet small and insignificant among the nations of the world as Israel was, without political influence or extended power, it may safely be said that no other people of antiquity holds a place of such profound importance in the history of human thought. It was Israel who gave to the world a religion which has directed the spiritual life of nearly half mankind, and, not only among the Jews themselves, but in the two daughter faiths of Christianity and Islam, has molded the beliefs of men in every continent save central and eastern Asia".[87]

William Oscar Emil Osterley and T.H. Robinson were historians who wrote the book *A History of Israel* published by Oxford University Press in 1932. The other aspect of this plague is the fact that it occurred in Palestine, which Shrewsbury referred to as the "cock pit"[88]) of the ancient world. According to Osterley and Robinson "Palestine was the commercial, military and political center of the ancient world, and on it focused all the greatest movements of the peoples"[89] because any conqueror had to control it if they wanted free movement from Europe or Asia into Africa or in the reverse. So controlling Palestine was vital because of this movement factor between Europe, Asia and Africa, the same factor that would also have easily spread the plague via the fleeing masses into adjoining regions thus spreading it throughout the Near East during the end of the Bronze Age.

Tularemia

The following hypothesis about Tularemia is included for the sake of completeness, so that no option is left out or eliminated from the list of potential infectious diseases that caused the end of the Near Eastern Bronze Age.

Tularemia, also known as Rabbit Fever, is a bacterial infection caused by *Francisella tularensis*. It is easily spread and is highly virulent making it the perfect infective agent for an epidemic. For these reasons it, along with dysentery (already discussed), the bubonic plague (already discussed), smallpox (to be discussed as the next infective disease), and anthrax (to be discussed later), have been used as infective disease agents for biological warfare.

Francisella tularensis infects small mammals such as rabbits, hares and other rodents who can in turn infect humans directly or via their ticks or mosquito or flies. "Humans can get the disease through a bite from an infected tick, horsefly, or mosquito; breathing in infected dirt or plant material (causing pneumonia); direct contact through a break in the skin with an infected animal or its dead body (most often a rabbit, muskrat, beaver, or squirrel) or eating infected rare meat".[90] So the disease is easily contracted either by direct contact with host rodents, ticks, mosquitoes, flies, ingestion of contaminated food and water or as aerosol particles being inhaled.

Siro Trevisanato, an Italian-Canadian molecular biologist proposed that tularemia was used by the Hittites as a germ warfare agent, caused the

Hittite Plague of 1322 BCE and later still caused the Plague of the Philistines. Trevisanato (1960–) graduated in 1983 with a BA in biology from SUNY Purchase, USA; in 1986 with an MSc in biochemistry from New York Medical College, Valhalla, USA and a PhD in 1994 in molecular biology from the University Copenhagen, Denmark.

Avaris was a port city in the northeastern region of the Nile Delta. It was the major administrative hub for the Nile Delta during the Hyksos era (Fifteenth Dynasty) and later became the capital of Egypt under the Hyksos. It was occupied from c.1783 to 1550 BCE and according to Trevisanato was the port through which tularemia entered Egypt c.1715 BCE.

Paragraph 170, column 11, lines 12 to 15 of the Hearst Papyrus tells of an infectious disease attributed to "the Asiatics", that is, people from the Syria—Canaan—Transjordan area also known as the "Canaanite Illness".[91] It has also been attributed to bubonic plague or typhus.

> Austrian archaeologist Manfred Bietak uncovered the ruins of Avaris at Tell el-Dab'a in the Eastern Nile Delta. There he also found unequivocal signs of an epidemic. More precisely, the layer dividing the so-called stratum G from stratum F of Avaris contains a large number of tombs, which are characterized by a hasty job. This fact is the signature of emergency situations such as those existing during a plague. Artefacts in the tombs revealed that they and, thus the epidemic, could be dated to c.1715 BCE.[92]

During the epidemic of 1715 BCE the Hebrew population in Avaris fared a lot better than the local Egyptians because the Egyptians living in the city had no immunity to tularemia while the Hebrews, who were exposed to sheep and other animals, had a natural immunity to tularemia and survived. If the disease had been bubonic plague or typhus then the Hebrews would have died at the same rate as the Egyptians. Trevisanato also thinks that the Hittite Epidemic of 1322 BCE was due to tularemia rather than bubonic plague.

> A deadly epidemic, also dubbed the Hittite plague, affected most of the Middle East towards the end of the fourteenth century BCE. The present study determined that its onset was described in the Egyptian royal archives, enabling to date it to the last reigning years of Akhenaten, that is, just before 1335 BCE, and locate the focus to an area northeast of Byblos (present-day Lebanon).

> Letter EA 96 states that "there is a pestilence in Simyra". Anyone from Simyra was barred from entering nearby Byblos, and donkeys were not to be used in caravans because of the pestilence. The measure did not work, as evidenced by letter EA 362 stating the pestilence did reach Byblos, and by letter EA 137 stating the Byblos ruler became chronically ill. The plague spread further south as attested by the ruler of Amurru in present-day southern Lebanon, who referred to his relationship with Byblos, and mentioned he was now sick, and was going to die (EA 95). Still further south, along the coastal trade route from Byblos, coeval letter EA 224 reports Megiddo "is consumed by pestilence".
>
> The east-west trade road going through Simyra linked the Mediterranean coast to the Euphrates. Reports from 1335 BCE show that east of Simyra in Babylon, an aristocratic woman died from plague (EA 7), and the local ruler was ill (EA 11), consistent with the spread along the trade route. West of Simyra, coeval letter EA 35 from Cyprus stated "the hand of Nergal is in my country", killing many, in particular individuals linked to copper mining. The attribution of the plague to the Mesopotamian god of pestilence Nergal points to an origin from the east, that is, via Canaanite harbors and indicates that the etiological agent also travelled by ship.[93]

Note that the letters referred to in the above quote are Amarna letters and are quoted from William Moran's book *The Amarna Letters* published in 1992 by the John Hopkins University Press, USA.

Trevisonato argues that initially the epidemic was confined to the central area of the Near East from Cyprus to Iraq and from Israel to Syria, sparing Egypt and the Neshite or Hittite Empire. Later, after fighting between Egypt and the Hittites at Amka on the Litani River to east of Byblos and Simyra in c.1325 BCE, the tularemia infection spread into Anatolia via the Egyptian prisoners of war to cause the 20-year Hittite Plague. It would have also spread via the various trade routes that went through the Levant.

In c.1320 BCE, Arzawa in western Anatolia attacked the Hittites. Eventually, after a Hittite counterattack, the Hittites laid siege to the capital of Arzawa. During this siege the Hittites left sheep carrying tularemia outside the city walls. The starving inhabitants of the city brought the sheep into their city where they, unbeknown to the city inhabitants, spread the disease among the population. Thus this is the first documented example of biological warfare.

Trevisonato argues why he thinks the epidemic was caused by tularemia:

> The reconstruction of the dynamics of the epidemic helps identifying the etiological agent. A disease lasting 35–40 years, infecting humans and animals, causing fever, disabilities, and death, spreading via rodents aboard ships as well as donkeys, points to *Francisella tularensis*, the etiological agent of tularemia. This disease can linger for a long time, and its longevity is incompatible with shorter-lived epidemics such as from bubonic plague, which for instance hit Europe around 1347–1349, and in 1629–1634. Furthermore, the description in Neshite records, e.g. knees, debilitation, and sensation of internal burning, is also coherent with tularemia. Moreover, tularemia also fits the onset of the infection, as it infects caravans stopping for rest, turning them into carriers for the etiological agent. The aforementioned trading route between the Mediterranean Sea and the Euphrates would have had such contact points allowing for the spread of the pathogen.[94]

Trevisonato further believes that the Plague of the Philistines was also due to tularemia, which has Canaan, the setting for the Plague of the Philistines, as a natural reservoir as shown above. He does not believe it was due to bubonic plague or dysentery and puts forward a strong argument:

> An etiological agent for the Philistine plague. The biblical data appear to center around the box as a vehicle for the disease, as well as the rodents that appear shortly thereafter, and are depicted in the "settlement" paid in gold. The Hebrew word *akhbar* for the rodents fails to distinguish between mice and rats. Rats would have carried *Y. pestis* but bubonic plague fails to adequately explain the epidemic. Mice are a better option; they can carry diseases, and fit the other data relative to the historical text, i.e. box, idol, and settlement payment.
>
> The gold-plated wooden box measured 2.5 × 1.5 × 1.5 cubits (Ex. 25.10-22; Ex. 37,5-10), that is, 1.1 × 0.7 × 0.7m, giving a volume of roughly 500 l, offering a nest to mice but not rats. The former animals average 20g and are small enough to enter the box through a small aperture possibly hidden by the gold covering. The latter animals average 300g requiring a wider aperture and more internal space. Mice nesting in the box would have explored their new habitat upon each transfer of the box, thus offering an explanation for the box transmitting the disease.
>
> Mice also explain the otherwise odd detail of a small Philistine idol falling on the floor. Once the box was hosted in the Philistine temple, the animals exiting the box from the same aperture, would have tipped over the statuette, eventually breaking the extremities after repeated falls (1Sa.5.2-5).

> The five replicas in gold of rodents and tumors to settle the dispute with the Hebrews (1Sa.6.3-5) also favor mice over rats. Given the specific gravity of gold, just over 19 kg/l, a gold mouse would be shy of 400 g, while a rat would be shy of 6 kg. Considering 10–20 g tumors, the Philistines were paying roughly 3–4 kg of gold in total. Rat-like tumors would have resulted in 31–32 kg of gold, where the tumors would have only contributed marginally (additional 3–6%) to the gold already provided by the rats, raising the question of their raison d'être.
>
> Linking mice to the box and to the disease singles out tularemia as the disease portrayed by the biblical text; mice are known to carry *Francisella tularensis*, the etiological agent for tularemia. Moreover, the text calls for a disease that originated from animals, can be communicated, can form tumors, and is deadly.[95]

So now there are arguments for the three virulent infections namely bubonic plague, dysentery and now tularemia, being present in the Near East at the end of the Bronze Age.

Smallpox

Smallpox or variole is a virulent infection caused by a virus. It was first called smallpox in Europe in the late fifteenth century to distinguish it from the "great pox" or syphilis. Thought to have originated in north eastern Africa 10,000 BCE, smallpox then spread to Egypt and from there onto India.[96]

Ramesses V is the fourth pharaoh of the Twentieth Dynasty of Egypt and he died c.1157 BCE[97] most likely from smallpox. The well-preserved mummy of Ramesses V shows classical smallpox lesions on the face, neck and shoulders as verified by Donald R. Hopkins in 1979.[98] This period was a time of expansion of the Egyptian Empire, so cases of smallpox could have been imported into Egypt because of this expansion into new territories and war. If the pharaoh, who would have been protected from all harm, finally succumbed to smallpox; how long had it been ravaging the general population of Egypt and how far had it spread in the Near East?

Sir Marc Armand Ruffer in 1910 described a smallpox like rash on a mummy from the same period (Twentieth Dynasty) as Ramesses V,[99] thus giving further primary physical evidence of the existence of smallpox in the Near East at the time of its ending between 1200 and 1150 BCE. Sir Marc

Armand Ruffer (1859–1917) was an Anglo-French pathologist who pioneered paleopathology. In 1882 he graduated from Oxford with a BA, then in 1887 gained his MBChB from University College in London followed by his MD in 1889. In 1916 he was knighted for his services to bacteriology and hygiene and for his services to the Red Cross Organization.

Normally a pharaoh is mummified and buried precisely 70 days into the reign of his successor[100] but Ramesses V was buried two years after his death—why? Hopkins suggests three possible reasons for this. First the body may have deteriorated due to the infection hence it needed prolonged mummification. Second the embalmers feared being infected by the smallpox. Finally there may have been a shortage of embalmers because they too had been killed by the smallpox epidemic. There may have also been a shortage of stone masons and stone cutters as well, because Ramesses VI was also buried in the tomb of Ramesses V, that is, two pharaohs in the one tomb, which is not the usual practice.

Tom Slattery has degrees in East Asian Studies from the University of California, Berkeley and in English from Central Washington University. In his book *The Tragic End of the Bronze Age. A Virus makes History*[101] Slattery argues that smallpox may have killed Ramesses V and started the end of the Bronze Age. He also argues that as people died with smallpox there were not enough men to mine tin that was vital in bronze production, hence the Bronze Age ended because less bronze was able to be made. Slattery also thinks that the Hittite Plague of 1322 BCE was due to smallpox.

Tuberculosis

The evidence for the existence of tuberculosis during the Bronze Age comes again from Sir Marc Ruffer and his examination of Egyptian mummies. In his book *Studies in the Paleo pathology of Egypt* he identifies the typical spinal lesions of Pott's disease as shown in Plate IX, figures 14 and 15.[102] Modern science has also identified the DNA of *Mycobacterium tuberculosis* in the spine of Egyptian mummies,[103] thus providing primary physical evidence for the existence of tuberculosis in the Near East in the late Bronze Age. In the crowded living conditions in cities with malnutrition and poverty, tuberculosis would have had the perfect breeding conditions in which to flourish and devastate the local population in the late Bronze Age with its high mortality rate.

Influenza

Influenza in an unprotected population had the potential to be catastrophic but there is no good evidence to support this theory. We only have its potential as a cause of an epidemic.

Poliomyelitis

Poliomyelitis had the potential to be devastating and Egyptian paintings confirm its existence in the late Bronze Age, so it is a definite possible cause of an epidemic.

Anthrax

Anthrax is a bacterial infection caused by the bacillus anthracis and is usually fatal in both humans and animals. Because it is so lethal it is used as one of the agents in biological or germ warfare today. In the late Bronze Age with no vaccine or antibiotics an epidemic would be catastrophic.

The endospores of the bacterium anthracis can survive for decades or even centuries and when they are inhaled, ingested or come in contact with the skin they can cause the disease in animals or humans. Herbivorous animals usually ingest the spores whilst grazing and carnivores usually ingest the spores when eating an infected animal. Humans usually contract the disease by inhalation of spores, direct contact of spores with the skin or consumption of the flesh of an infected animal.

One of the Ten Plagues suffered by Egypt, as described in the Bible, could have been anthrax with the description of the typical anthrax sores on the animals, which confirms that anthrax existed in the Near East in the Bronze Age; hence this is another possible cause of an epidemic.

Measles

Measles in a vulnerable population has the potential to cause a lethal epidemic, but unfortunately there is no good evidence or records to show it occurred. We only have its potential as a cause of an epidemic.

Malaria

Malaria has been infecting humans "for the entire history of the species"[104] and existed in the Near East from Anatolia through the Levant and into the Nile in Egypt in the late Bronze Age, as it still does today. So it had the constant potential to cause death in humans, especially the virulent Falciparum strain during the end of the Bronze Age era.

Typhus

Typhus is a zoonosis that is transferred to humans via lice. It was common among groups of people who lived very closely together such as soldiers in barracks, sailors in ships, prisoners in gaols and refugees in camps; hence it was known variously as "barrack fever", "ship fever", "gaol fever" or "camp fever".

Was the epidemic that the Egyptian prisoners of war took into the Hittite Empire, that caused the Hittite Epidemic of 1322 BCE, really typhus "camp fever" and not smallpox or bubonic plague ? Also, as discussed earlier—typhus may have caused the Plague of the Philistines.

In summary, what was needed was an epidemic or several epidemics that had the ability to be lethal, widespread and able to continue for long periods of time or recur frequently; this would be able to change the course of history by destroying empires such as the Mycenaean and Hittite Empires and ending the Bronze Age. Because the end of the Bronze Age is approximately 3,200 years ago, then the virgin population factor has to be taken into consideration as some of the epidemics, such as influenza and measles, may have been the first appearance of these infections in history, hence they would be a lot more lethal than they are today in the same area, the Near East.

Of the diseases discussed above, there is evidence that bubonic plague, smallpox and tularemia fit these criteria and there is evidence of their existence in the Near East at the time. Dysentery and typhus would be more localized and shorter lived, but still very lethal in its time and place.

Poliomyelitis and tuberculosis would be too slow to kill and spread over a wide area but potentially lethal locally. Anthrax and malaria are also possibilities. There is no good evidence or records to support the existence of influenza and measles. They may well have occurred and been lethal, but if they did occur they would have been seasonal and relatively short-lived epidemics.

This is in agreement with Itamar Singer's understanding who said "plagues are already attested in Anatolia in the Old Assyrian Colony period (Cecen 1995), and are often mentioned in Late Bronze Age Syrian documents (Klengel 1999b)".[105]

So it is possible that diseases could have contributed to the end of the Hittite Empire and the end of the Bronze Age in a major way. Because of the timeframe of about 50 years from c.1200–1150 BCE there was plenty of time to have several major lethal epidemics such as bubonic plague, smallpox and tularemia occur on a widespread scale, supported by more local epidemics such as dysentery, poliomyelitis, tuberculosis, measles, influenza, anthrax and malaria. In a time of poor hygiene, no antibiotics and no vaccines these infections (five of which are so virulent that they are still used as germ warfare agents today, namely bubonic plague, smallpox, tuberculosis, anthrax and dysentery) had the potential to devastate the Near East and should be factored in as potential cofactors in the future by historians. Any future discussions about the end of the Hittite Empire and end of the Bronze Age should be considered as incomplete if disease is not considered and included.

Notes

1. Pernicka, E. et al., "*Early Bronze Age Metallurgy in Northeast Aegean*", in Wagner, G.A. et al.(eds), *Troia and Troad: scientific approaches, natural science in archaeology*, (Berlin, London: Springer, 2003) 143–172.
2. Cowen, R., "*The Age of Iron*", chapter 5 in a series of essays on *Geology, History and People* prepared for a course at the University of California at Davis.
3. Danzer, G.A., *Atlas of World History*, (Kent, UK: Grange Books, Rochester, 2002), 23 and "*Ironware piece unearthed from Turkey found to be the earliest steel*", (The Hindu, March 26, 2009).
4. Soggs, H.W.F., *The Babylonians*, (London: The Folio Society, 2005), 415.
5. Roberts, J.M., *Ancient History from the First Civilization to the Renaissance*, (London: Duncan Baird Publishers, 2004), 69.
6. Waldbaum, J.C., *From Bronze to Iron. Studies in Mediterranean Archaeology*, (Goteburg: Paul Astroms Forlag, 1978), 56–58.
7. Snodgrass, A.M., *Arms and Armour of the Greeks*, (London: Thames and Hudson, 1967) and Snodgrass, A.M., *The Dark Age of Greece*, (Edinburgh: Edinburgh University Press, 1971).
8. Cowen, R., *The Age of Iron*, online version of the course on http://mygeologypage.ucdavis.edu/cowen/GEC115/115CH5.html. Accessed 3 December 2013.
9. *Collapse of the Bronze Age*, http://teachingcompany.12.forunner.com/a/9-the-collapse-of-the-bronze-age_post2466.html, 1. Accessed 30 October 2011 and "*The Catastrophe*", http://www.brusselsjournal.com/node/4106, "*The Catastrophe*" *Part 2: What the End of the Bronze-Age Civilization means for Modern Times*, 1. Accessed 25 September 2011.
10. Ibid, "*The Catastrophe*".

11. *Bronze Age Collapse*, http://www.enotes.com/topic/Bronze_Age_collapse, 1. Accessed 30 October 2011.
12. Lesson 28, *The Collapse of Mycenaean Palatial Civilization*, http://projectsx.dartmouth.edu/history/bronze_age/lessons/les/28.html, 15. Accessed 25 September 2011.
13. *The End of the Bronze Age*, http://www.mcgoodwin.net/pages/otherbooks/rd_endbronzeage.html, 16. Accessed 25 September 2011.
14. Ibid, http://www.enotes.com/topic/Bronze_Age_collapse, 3–4. Accessed 30 October 2011.
15. Cunliff, B., *Iron Age Communities in Britain* (*4th Ed.*), (Routledge, 2005) 256 and Yurco, G.J.,"*End of the Late Bronze Age and Other Crisis Periods*: *A Volcanic Cause*" in Teeter, E. and Larson, J (Eds) *Gold of Praise*: *Studies on Ancient Egypt in Honor of Edward F. Wente*. (*Studies in Ancient Oriental Civilizations*) (Chicago: Oriental Institute of the University of Chicago, 1999) 456–458.
16. Baillie, M.G.L., "*Hints that Cometary Debris Played some Role in Several Tree-Ring Dated Environmental Downturns in the Bronze Age*" in Peiser, B.J., Palmer, T. and Bailey, M.E., *Natural Catastrophes During Bronze Age Civilizations. Archaeological, Geological, Astronomical and Cultural Perspectives*. (Oxford: Bar International Series 728, Archaeopress, 1998) 109.
17. Ibid, 110.
18. Ibid, 109–116.
19. Nur, A. and Cline, E., "*Poseidon's Horses*: *Plate Tectonics and Earthquake Storms in the late Bronze Age Aegean and Eastern Mediterranean*", (Journal of Archaeological Science, No. 27, 2000) 43–63.
20. Peiser, B.J. op.cit. "*The End of the Bronze Age by Large Earthquakes*?" 140–147.
21. Ibid, 140.
22. Ibid, 143.
23. Ibid, 145.
24. Ibid, 146 and French, E.B., "*Evidence for an Earthquake at Mycenae in Archaeoseismology*", S.B. Stiros (Ed), (Exeter, Great Britain: The Short Pan Press, 1996) 51–54.
25. Schachermeyr, G., *Griechische Fruhgeschichte*, Verlag, der Osterreichischen Akademie der Wissenschaften, 1984. Wien. (Vienna: Greek Prehistory, the Austrian Academy of Sciences, 1984).
26. *Bronze Age Collapse*, http://www.enotes.com/topic/Bronze_Age_collapse, 3. Accessed 30 October 2011.
27. Gardiner, A.H., (1947), *Ancient Egyptian Onomastica*, (London: Oxford University Press, Vol 1,1947) 196. And Mannassa, C., "*The Great Karnak Inscription of Merneptah*: *Grand Strategy in the Thirteenth Century BC*". (New Haven: Yale Egyptological Seminar, Department of Near Eastern Languages and Civilizations, Yale University, 2003) 55.

28. Pritchard, J.B.(Ed.), *Ancient Near Eastern Texts relating to the Old Testament, 3rd Edition*, (Medinet Halu inscriptions of Ramesses III's eighth year lines 16–17 as translated by John A. Wilson, 1969) 262.
29. Nougayril, J. et.al. *Ugaritica*, (Vol: 87–90, 1968) no. 24.
30. *Who were the Sea People*? http://www.saudiaramcoworld.com/issue/199503/who.were.the.sea.people.htm. Accessed 3 December 2013.
31. Herodotus, 1.94.
32. Grant, M., *The Ancient Mediterranean*, (New York: Charles Scribner's Sons, 1969) 79.
33. Ibid and Finlay, M.I., *Early Greece: The Bronze and Archaic Ages, New and Revised Edition*, (London: W.W.Norton & Co, 1981).
34. Refer to: Sorenson, M.L.S. and Thomas, R. (eds), *The Bronze Age – Iron Age Transition in Europe*, (Oxford University Press, 1989) and Wertime, T.H. and Muhy, J.D. *The Coming of the Age of Iron*, (New Haven Publishers, 1980).
35. Herodotus, *New Oxford American Dictionary*, (Oxford University Press).
36. Topal, H., "*Contribution of drought to the collapse of the Hittite Empire*" on http://ie499.yeralan.org/index.php?option=com_content&view=article&id=322:hititler&catid=53:projects&ltemid=61, (2009) 2. Accessed 18 September 2011.
37. Kuniholm, P.I., Newton, M., Griggs, L.B. and Sullivan, P.J., "*Dendrochronological Dating in Anatolia: The Second Millenium BC*", (der Anschnitt, Anatolian/metal III, Beiheff 18, 2005) 41–47.
38. Drake, B., "*The influence of climatic change on the Late Bronze Age Collapse and the Greek Dark Ages*", (Journal of Archaeological Science, XXX, 2012) 1–9.
39. Moody, J., "*Changes in Vernacular Architecture and Climate at the end of the Aegean Bronze Age*", in *Forces of Transformation – The end of the Bronze Age in the Mediterranean*, (Oxford: Oxbow Books, 2009) 6–19.
40. Ibid, http://projectsx.dartmouth.edu/history/bronze_age/lessons/les/28.html, 14. Accessed 25 September 2011.
41. Ibid.
42. Ibid, http://mcgoodwin.net/pages/otherbooks/rd_endbronzeage.html, 2. Accessed, 25 September 2011.
43. Ibid.
44. Tainter, J., *The Collapse of Complex Societies*, (Cambridge: Cambridge University Press, 1976).
45. Vermule 1960.
46. Lokovides 1974.
47. Betancourt 1976.
48. Stiebing, (op.cit.) 231.
49. Mieroop, (op.cit.) 198.

50. Dickinson, (op.cit., 2010) 489.
51. Dickinson, (op.cit. 2006) 56.
52. Middleton, (op.cit.) 165.
53. Ibid, 166.
54. Ibid.
55. Ibid.
56. Ibid, 164.
57. King, H., *Health in Antiquity*, (Oxford: Routledge, Abingdon, 2005) Chap. 1.
58. Ibid, 21.
59. Ibid, 30.
60. Watson-Williams, (op.cit.) 110.
61. Ibid, 123.
62. Walloe, (op.cit.) 122.
63. Ibid.
64. Ibid.
65. Snodgrass, M.E., *World Epidemics. A Cultural Chronology of Disease from Prehistory to the Era of SARS*, (North Carolina: McFarland and Co. Inc., Jefferson, 2003) 20.
66. Achtman, M. et al., "*Yersinia Pestis genome sequencing identifies patterns of global phylogenic diversity*", (Native Genetics, 42, 2010) 1143–1150.
67. Ibid.
68. Ehrlich, C.S., *The Philistine in Transition; A History from ca. 1000 – 730 BC*, (Leiden, The Netherlands, 1996).
69. Riemschneider, M., "*Die Herkunft Der Philister*", (Acta Antiqua IV, 1956) 17–29.
70. Underwood, E.A., Book Reviews "*The Plague of the Philistines, and other Medical – Historical Essays by J.F.D. Shrewsbury*", (London; Victor Gallancz, 1964), 189 in (British Journal for the History of Science, vol. 2, no. 2, 1964) 169–171.
71. Snodgrass, M.E., (op cit), 12 and Kohn, G.C. (ed), *Encyclopedia of Plague and Pestilence from Ancient Times to the Present*, (New York, USA: Checkmark Books, 2001) 266–67.
72. Kuniholm, P.I., "*Anatolian Tree Rings and a New Chronology for the East Mediterranean Bronze – Iron Age*", (Science, 21 December 2001, Vol. 294, no. 5551) 2532–2535 and Kuniholm, P.I., "*Anatolian Tree Rings and the Absolute Chronology of the Eastern Mediterranean 2220 – 718 BC*", (Nature, Vol. 381, 27 June 1996) 782.
73. Davis L.A., *Natural Disasters*, (Facts on File Science Library, 2008) 219.
74. Gregg, C.T., *Plague an Ancient Disease in the Twentieth Century*, (Revised Edition, Albuquerque: University of New Mexico Press, 1985).
75. Ibid, 7.

76. Simpson 1905.
77. Topley and Wilson 1946.
78. Shrewsbury, J.F.D., "*The Plague of the Philistines*", (London: I. Hyg, September, 47 (3), 1949) 245.
79. Watson-Williams, E., (op.cit.).
80. Walloe, L., (op.cit.).
81. Shrewsbury, J.F.D., (op. cit., 1949) 244–252.
82. Ibid, 250.
83. Whiston, W., *The Works of Flavius Josephus translated by W. Whiston*, (London, 1793).
84. Shrewsbury, J.F.D., (op.cit, 1949) 250–251.
85. Ibid, 249–250.
86. Ibid, 250.
87. Ibid, 251.
88. Ibid, 244.
89. Osterley, W.O.E. and Robinson, T.H., *A History of Israel*, (Oxford: Oxford University Press, 1932).
90. Tularemia, http://www.ncbi.nlm.nih.gov/pubmedhealth/PMH0001859/, 1. Accessed 3 December 2013.
91. Trevistanato, S. I., "*Did an epidemic of Tularemia in Ancient Egypt affect the course of world history*", (Elsevier Ltd, Medical Hypotheses, Vol. 63, 2004) 905.
92. Ibid, 906.
93. Trevisanato, S.I., "*The "Hittite Plague", or epidemic of tularemia and the first record of biological warfare*", (Elsevier, Medical Hypotheses, Vol. 69, 2007) 1371 and Moran, W.C., *The Amarna Letters*, (Baltimore: John Hopkins University Press, 1992).
94. Trevisanato, (op.cit, 2007) 1373.
95. Trevisanato, "*The biblical plague of the Philistines now has a name, tularemia*", (Elsevier, *Medical Hypotheses*, Vol. 69, 2007) 1145.
96. Smallpox, http://www.infoplease.com/cig/dangerous-disease-epidemics/smallpox-12,000-years-terro. Epidemics of the Past Smallpox: 1200 years of Terror, 1. Accessed 24 September 2011.
97. Snodgrass, M.E., (op.cit.) 12 and smallpox, http://www.gorydetails.net/demo_sites/Smallpoxsite/smpx_history.ol.html, 1. Accessed 7 December 2013.
98. Whqlibdoc. Who.int/smallpox/WH_5_1980_p22.pdf. *Ramesses V. Earliest known victim*, Donald R. Hopkins, 22.
99. Ruffer, M.A. *Note on an Egyptian resembling that of Variola in the Skin of a Mummy of the Twentieth Dynasty 1200–1100 BC*, (Journal of Pathology and Bacteriology, Vol. 15, 1910) 1–4.

100. Clayton, P., *Chronology of the Pharaohs*, (Thames and Hudson Ltd., 1994) 167.
101. Slattery, J., *The Tragic End of the Bronze Age A virus makes history*, (New England, USA: Writers Club Press, Lincoln, 2000).
102. Ruffer, M.A., Moodie, R.L. (Ed), *Studies in the Paleopathology of Egypt*, (Illinois: The University of Chicago Press, Chicago, 1921).
103. Zink, A, et al., "*Characterization of Mycobacterium Tuberculosis Complex DNAs from Egyptian Mummies by Spoligotyping*", (Journal of Clinical Microbiology, Vol. 41(1), 2003) 359-367.
104. Joy, D., Feng, X, et al., "*Early Origin and Recent Expansion of Plasmodium Falciparum*", (Science, Vol. 300 (5167) 2003) 318–321.
105. Singer, I., (op.cit). 47 and Cecen, S., "*Mutanuinden Kultepe-Texten*" in (Atti del 11 Congresso Internazionale di Hittitologia, 1995) 101–102. O. Carruba, M. Giorgieri, C. Mora. Eds. Pavia: Gianni Iuculano and Klengel, H. (1999b) "*Epidermien im spatbronzezeitlichen Syrien – Palastina*" in Michael. (Historical, Epigraphical and Biblical Studies in Honor of Michael Heltzer. Ed. Y. Avishur and R. Deutsch. Tel-Aviv-Jaffa: Archaeological Center Publications) 187–193.

CHAPTER 6

How Disease Affected the History of Ancient Carthage

Carthage was founded in 814 BCE by Queen Dido, who had been exiled from the Phoenician (or Punic) city of Tyre,[1] which was situated on the east coast of the Mediterranean. Carthage was located on the eastern side of Lake Tunis, across from the center of today's Tunis in Tunisia, on the North African coast. It was to become one of the largest cities in Greek times because of its strategic location. All shipping trade routes in the Western Mediterranean had to pass between Sicily and the coast of Tunisia, so control of this passage meant control of the western Mediterranean.

The city was built on a promontory jutting out into the Mediterranean Sea, which gave it access to the sea via the northern and south eastern shores; plus it built two large artificial harbors within the walled city—one for merchant ships, known as the Mercantile Harbor, and one for its large navy of up to 350 warships at its peak, known as the War Harbor. The whole city complex was surrounded by huge walls 37 kilometers long.

Carthage became rich and powerful more through agriculture, (growing wheat, olives and grapes), trade and alliances than by war, conquest and garrisons. It eventually took over the main Phoenician cities of Tyre and Sidon in power, influence and wealth. The Carthaginian Empire dominated the western Mediterranean extending from southern Spain across the top of North Africa into Corsica, Sicily and Italy.

Despite trying to develop its influence through peaceful means Carthage had two main rivalries, which eventually led to war. One was with Sicily and its main town of Syracuse, which was the largest seaport in Sicily. Sicily

P. Norrie, *A History of Disease in Ancient Times*,
DOI 10.1007/978-3-319-28937-3_6

Image 6.1 Western Mediterranean & Carthage (© Hemera Technologies/ AbleStock.com/Thinkstock)

Image 6.2 Photograph of Ancient Carthage Ruins (© derejeb/iStock/ Thinkstock)

was important because it controlled the other side of the passageway that ships had to traverse to get to the western Mediterranean, so Carthage could then completely control this trade route, if it also controlled Sicily. Syracuse naturally did not want to lose this control to Carthage so many wars were fought between Carthage and Syracuse. If Carthage won these wars it would completely control the western and central Mediterranean and then be able to expand into the eastern Mediterranean eventually controlling the whole Mediterranean, as Rome would do later. So control of Sicily was the key to Carthage's destiny.

The other main rivalry Carthage had was with a new Italian city state, just starting to expand, called Rome. Eventually Carthage and Rome fought three great wars called the Punic Wars, which determined who was going to control the Mediterranean in the post-Greek Empire era. If Carthage had won Rome would have been stopped before it really got started; so there would not have been a Roman Empire at all. If Rome had won, the Carthaginian Empire would have ended at its peak. Eventually Rome won and after the Third Punic War Rome took its revenge by completely destroying Carthage. "The call of Cato–'Carthage must be destroyed'–was to be famous as an expression of an implacable hostility arising from fear".[2] Rome did not want to leave any possibility that Carthage could ever re-emerge as a rival so it was laid to waste, totally destroyed and burnt to the ground. It was as if Carthage never existed and that is why today there original Carthaginian records are few and far between—they had all been destroyed by Rome. This total destruction of Carthage in 146 BCE was the savage and barbaric act of a jealous rival and not worthy of the civilized ideals of Rome. The Roman General Scipio Aemilianus also burnt all the Phoenician warships in the War Harbor and sold 50,000 Carthaginians, those who had not been killed, into slavery.

Historians such as Roberts, Blainey and McNeill just state that Rome followed Cato's cry and destroyed Carthage, razing it to the ground; but none of them mention any diseases as contributing cofactors in the fall of Carthage. Diamond and Watts don't mention Carthage or its fall at all.

Disease played its role in the fall of Carthage, as the following four examples will show. If it was not for major outbreaks of infection Carthage could have defeated Syracuse and Rome and been the dominant force in Sicily and the Mediterranean. Rome may never have been. Instead it was Carthage that fell.

The first three examples deal with Carthage's war with Syracuse for control of Sicily and thus the middle and western Mediterranean. Sicily was the key to controlling the central Mediterranean Sea and the two straits on either side of it controlled access to the western Mediterranean

Sea. So Sicily was vital to Carthage's long-term plans and to its eventual survival. The fourth example deals with Carthage's ongoing Punic Wars with Rome.

The first Punic War was between 264 and 241 BCE and resulted in Sardinia and western Sicily being taken from Carthage and becoming Roman provinces. The second Punic War lasted from 218 to 201 BCE. Here the famous Carthaginian General Hannibal crossed the Alps with his army, which included elephants, to invade northern Italy and thus attack the heart of Rome directly. This army brought malaria with them,[3] which infected the Romans thus helping Hannibal defeat them at the battles of Lake Trasimere in 217 BCE and Cannae in 216 BCE.

At Cannae Hannibal destroyed a Roman Army twice the size of his army—something that threatened Rome's very existence. Hannibal, however, was now far from his home base, depleted in numbers after many battles and campaigning in a barren countryside, which was not able to contribute much in the way of supplies. So he was not able to take the city of Rome itself. When Rome boldly started attacking Carthaginian colonies in Spain and North Africa under the leadership of the gifted commander Cornelius Scipio, Hannibal had to abandon his north Italy campaign to meet these new Roman threats and was finally defeated at the battle of Zama in 202 BCE. The third Punic War dated from 149 to 146 BCE and resulted in the total destruction of the city of Carthage; Cato cried "Carthage must be destroyed"[4] because he feared that it would rise again to challenge Rome.

The Four Examples

First Example

The Carthaginian Empire in Northern Africa and Spain was trying to expand into Sicily. While General Himilco marched on the Sicilian seaport of Syracuse in 405 BCE his army was infected by a devastating disease, thought to be measles.[5] With his army nearly destroyed and his cousin dead, Himilco had to make peace with the Syracusan leader Dionysius the Elder and abandon most of his outposts in Sicily, except in the west. The Carthaginian soldiers took the disease back with them to Carthage where it spread among its citizens. In 397 BCE another outbreak occurred as Himilco again tried to conquer Dionysius the Elder of Syracuse. Again disease won the day and Himilco lost his war.

Second Example

In 396 BCE[6] the Greek historian Diodorus of Sicily described the symptoms of a plague that so weakened the Carthaginian Army and Navy in Sicily that Dionysius of Syracuse was able to destroy them and may have prevented Carthage from dominating the Eastern Mediterranean for many years to come. These symptoms included "Inflammation and mucous in the throat; burning, pain, and fatigue throughout the body, diarrhea, pustules on the skin and delirium"[7] and "pustules throughout the surface of the body as well as ... spine aches and mental symptoms".[8] US epidemiologist, Hans Zinsser thinks the symptoms are similar to those described by Thucydides in the Plague of Athens and may have been due to a severe type of smallpox infecting an unprotected population with appalling results and consequences.[9]

Donald R. Hopkins in his book on smallpox *The Greatest Killer Smallpox in History* only devotes one paragraph to the story of Carthage and its fall. But he comes to an excellent conclusion at the end of the paragraph with the following last sentence: "As a result of this epidemic, Carthage was unable to gain control of Sicily, which could have given it a decisive strategic advantage in its struggle with the fledgling Roman Empire during the Punic Wars less than a century later".[10]

Third Example

In the 344–339 BCE war with Timoleon of Syracuse[11] the Carthaginian Army was able to occupy all of the city of Syracuse except for the citadel. Again disease came to the rescue of the Syracusans in the form of a plague (type unknown), which so weakened the Carthaginian Army that Timoleon was able to drive them out of Syracuse and eventually defeat them at the Battle of the Crimissus in 340 BCE. The ensuring peace terms were unfavorable to Carthage.

Fourth Example

During the Second Punic War between Rome and Carthage much of the fighting centered on Sicily and the Carthaginian stronghold of Syracuse. In 212 BCE [12] the besieging Roman soldiers succumbed to a serious infection, most likely influenza as did the Carthaginians eventually. To avoid infection, Roman General Marcus Claudius Marcellus marched his forces, around the infected lowland areas and then captured Syracuse thus ending Carthage's control of Sicily.

The fall of Sicily was a major, if not fatal, loss to Carthage since it meant it no longer controlled the central Mediterranean and the passageway to the Western Mediterranean, which were now under Roman control. Eventually Rome destroyed Carthage totally in 146 BCE when Scipio burnt the city to the ground over six days.

The latest book about Carthage is *Carthage must be Destroyed–The Rise and Fall of an Ancient Civilization* by Cambridge University PhD ancient historian, Dr. Richard Miles,[13] who is now a senior lecturer in Ancient History at the University of Sydney. Miles argues that the main reason Rome defeated Carthage was Rome's ability to match Carthage's naval superiority. According to him Rome's new fleet led to Rome's eventual victory.

Carthage was renowned for its naval design and ability to manufacture its vessels quickly. It was the first to develop the Quadrireme, a much larger and more powerful vessel than the usual Trireme in service with all other navies, with four levels of rowers instead of three. It ruled the waves, much like the Royal Navy did in the days of Nelson with its undisputed command of the Mediterranean Sea. Eventually Carthaginian naval architects developed the even bigger Quinquereme with five levels of rowers.

In 261 BCE one of these vessels was captured when it went aground while on a raid on the western Italian coast. Roman shipwrights analyzed the ship and were eventually able to not only copy it in large numbers but also were able to improve on its design.

All this naval power would be useless if one did not control Sicily and hence the two straits on each side of her namely the narrow Strait of Messina to the north east between Sicily and the Italian peninsula and the broader Sicilian Channel to the south west between Sicily and Tunisia, where the City of Carthage was located. Whoever controlled Sicily and its many ports was able to control these two straits hence the central and western Mediterranean Sea. Disease at four different times denied Carthage taking complete control of Sicily and that is why it is so important to give disease its rightful place as the main cofactor in the fall of the Carthaginian Empire.

In his 521 page book Miles mentions disease three times; briefly in one sentence on each of the pages 125 and 126 and for half a page between pages 128 and 129 when he discusses how an outbreak of possibly typhus in 396 BCE helped Dionysius of Syracuse defeat the Carthaginians once again.

The following quote describes the symptoms of the outbreak of a possible typhus infection. "The plague began with catarrh; then came a swelling in the throat; gradually burning sensations ensued; pains in the sinews of the back and a heavy feeling in the limbs; then dysentery supervened and pustules upon the whole surface of the body. In most cases this was the course of the disease; but some became mad and totally lost their memory; they circulated through the camp, out of their mind, and struck at anyone that they met. In general, as it turned out, even help by physicians was of no avail, because of both the severity of the disease and the swiftness of death; for death came on the fifth day or on the sixth at the latest, amid such terrible tortures that all looked upon those who had fallen in war as blessed".[14]

But Miles does not give any weight (as demonstrated by the amount of space he devoted to the topic) to the idea that the defeat of the Carthaginians by disease and thus any control of Sicily was significant or important enough. This book disagrees with this view.

In summary disease had a great effect of on the History of Europe and its environs in ancient times. Disease helped destroy the Hittite Empire, otherwise their Empire, with its technical advantage of iron weapons, could have controlled all of the Near and Middle East plus the Mediterranean.

Image 6.3 Photograph of Archaeological Park Carthage (© Witold Ryka/iStock/Thinkstock)

Disease limited the spread of the Carthaginian Empire eastward in the Mediterranean at various times, thus allowing Rome to expand its Empire. In other words Rome could have been destroyed by Carthage in its beginning before it became a great Empire. Disease also helped the Carthaginian general Hannibal to occupy northern Italy, thus threatening the survival of Rome itself.

What is evident is that the major historians and scholars who to date have chartered the rise and fall of the Hittite, Assyrian, Persian, Greek, Carthaginian and Roman Empires, have generally framed their work simply in terms of military, economic and political factors. While they do mention diseases and plagues, they do so in passing, and do not give sufficient weight to these factors as shapers of the course of history. By contrast, what has been demonstrated in this chapter is that no history of any of these empires and indeed the whole history of the ancient world, can be considered to be adequate unless it takes into account the effects of diseases such as smallpox, malaria, typhoid fever, measles, typhus, encephalitis and dysentery on both military forces and civilian populations, as well as individuals such as, Alexander the Great and the Hittite Kings Suppiluliumas I and his son Arnuwandas II, at crucial moments in the history of these ancient empires.

Notes

1. www.ancient.eu/carthage/page1 accessed Sept. 17, 2015—online Ancient History Encyclopaedia.
2. Roberts, J.M. op.cit, (2003), 235.
3. Snodgrass, M.E., op.cit, 16.
4. Roberts J.M. op.cit, (2002), 372.
5. Snodgrass, M.E., op.cit, 16.
6. Ibid, 57.
7. Ibid.
8. Alivizatos, G.P. *The Early Smallpox Epidemics in Europe and the Athens Plague after Thucidides*, Athens, (1950), 348.
9. Zinsser, H., *Rats, Lice and History*, (New York: Bantam Books, 1960), 92.
10. Hopkins, D.R., op.cit, (2002), 20.
11. Dupy, R.E. and T.N., *The Harper Encyclopedia of Military History from 3,500BC to the Present*, (New York: Harper Collins Publishers Inc., 1993), 63.
12. Kohn, G.C., op.cit, 279.
13. Miles, R., *Carthage Must be Destroyed*, (London: Penguin Books, 2010).
14. Ibid, 128.

CHAPTER 7

Conclusion

This book has discussed the history of disease in ancient times. The purpose of this book was to show that disease was a significant contributing causal factor in history; especially in ancient times and reflects the fact that history is not just made up of dates of battles and the list of monarchs and when they reigned. Second it looked at what the possible diseases may have been and third it questioned the status quo such as whether the 1322 BCE Hittite epidemic ended the Hittite Empire.

This book has shown that disease can have a significant impact on major historic events, that is, events that are significant enough to change the course of history, such as the end of the Bronze Age and the Hittite Empire. It has shown that disease, in the form of several infectious disease epidemics, fits the medical model to explain three factors about the end of the Bronze Age, namely the short time frame of the Catastrophe, the mass migrations of not only the general population but also the "Sea Peoples" and the abandonment of cities during the Catastrophe.

The importance of the impact of disease on history has been championed by a few sociologists and historians. Warren Robertson from Drew University, New Jersey, USA stated in his book *Drought, Famine, Plague and Pestilence Ancient Israel's Understandings of and Responses to Natural Catastrophes* that disease and other natural catastrophes such as earthquakes and droughts "are external agents of change that shape history and culture".[1] While Prof. Pitirim Sorokin, the foundation Professor of Sociology at Harvard University in Boston, USA said in his book *Man*

P. Norrie, *A History of Disease in Ancient Times*,
DOI 10.1007/978-3-319-28937-3_7

and Society in Calamity that "for good or ill, calamities are unquestionably the supreme disrupters and transformers of social organization and institutions".[2] Finally Dr. William H. McNeill, Professor of History at the University of Chicago thinks that infectious diseases are "one of the fundamental parameters and determinants of human society".[3]

This book has also argued that if the end of the Hittite Empire was due to civil war or invasion by neighbors then why was the capital Hattusa abandoned over a short period of time and not destroyed or occupied by the new victors, as the archaeological evidence has shown.

This book has shown that the 1322 BCE Hittite epidemic was not the fatal blow that ended the Hittite Empire as portrayed by Prof. Wynn, Prof. Hare and Dr. Cooray and that it was not caused by smallpox, as suggested by them, but by bubonic plague. It has also listed ten possible diseases that could be implicated in the end of the Hittite Empire 120 years later, along with the end of the Bronze Age. Of those ten diseases, five are so lethal that they are still used as germ warfare agents. But, bubonic plague, tularemia and smallpox remain the main possible diseases implicated as the causes.

This book has presented a comprehensive list of sixteen possible causes for the end of the Bronze Age and shown that a multifactorial scenario ending with an increased risk of infectious disease was the most likely outcome and that how much weight is given to other cofactors depends on the country or region involved, for example, an increased risk of earthquake in Greece. The only other things that could have been ubiquitous across the whole region would have been post- volcanic or post-comet climate change, leading to drought then famine and finally infectious diseases. These diseases know no boundaries; hence they spread across the whole region.

This book has asked many questions: If wasn't for the 1322 BCE Hittite epidemic could the Hittite Empire have won the battle of Kadesh and subsequently invaded Egypt? What caused the mass migrations of the general population if not disease? What caused the mass migrations of the "Sea Peoples" if not disease? Why were cities abandoned if not for disease? And finally why are historians not focusing on disease as a possible cofactor in history, often a very significant cofactor as shown by this book? This book has also shown that infectious disease epidemics contributed to the fall of ancient Carthage and played a significant role for change in the history of the Mediterranean.

So what have we learnt from history over the millennia? We have learnt that nothing has changed when it comes to infectious disease. New plagues such as HIV/Aids, Ebola and SARS are treated in the same way as the original plagues, namely with isolation and quarantine. That new emerging and re-emerging infectious diseases such as antibiotic resistant tuberculosis are now appearing and show us how impotent we are when it comes to controlling infectious diseases. We have learned that new strains of super viruses made from the combination of human, pig and bird viruses, such as H1N1 Swine Flu will become more common in the future and have the capacity to become drug resistant, hence severely threatening mankind yet again; as did the infectious diseases that ended the Bronze Age and Hittite Empire, thousands of years ago.

We have also learnt that further investigation is needed, looking for more evidence of infectious diseases in ancient times. Archaeological digs should engage a paleoentomologist to look for the remains of bubonic plague rats and fleas for example, as bubonic plague was one of the main scourges of the ancient world. Paleoparasitologists should also be used to determine what parasites may have infected the skeleton. Paleoparasitologists could sample the soil around where the abdomen would have been, looking for the eggs of any parasites. Then all skeletons found should have DNA testing, if possible, to identify any infectious diseases in the body that may have contributed to its death, along with examination by a properly trained forensic paleoanthropologist.

Finally this book has shown that in history, the medical aspects such as infectious diseases, should be taken into consideration every time and given more weight as causal factors of events than is currently the case.

NOTES

1. Robertson, W., *Drought, Famine, Plague and Pestilence: Ancient Israel's understandings of and responses to natural catastrophes*, (New Jersey, USA: Gorgias Press, Piscataway, 2010), 15.
2. Sorokin, P, *Man and Society in Calamity*, (New Jersey, USA: Transaction Publishers, Piscataway, 2010) 121.
3. McNeill, W.H., *Plagues and Peoples*, (New Jersey, USA: Anchor Books, 1998) 2.

Plague Prayers of King Mursili II

No. 8 Mursili's Hymn and Prayer to the Sun-goddess of Arinna (CTH 376.A)

Invocation

§l (E i 1–10; A i 1'-5') [0 Sun-goddess of Arinna! A mighty and honored goddess are] you! Mursili, [the king, your servant,] sent me [(?)[2]] saying: "Go and say to my [lady, the Sun-goddess] of Arinna: 'I shall invoke the Sun-goddess of [Arinna], my personal [goddess] (lit. of my head). [Whether] you, [0 honored] Sun-goddess of Arinna, are above in heaven [among the gods], or in the sea, or gone to the mountains [...] to roam, or if you have gone to an enemy land [for battle], now let the sweet odor, the cedar and the oil summon you. Return to your] temple! [I am herewith invoking you] by means of offering bread [and libation]. [So] be pacified and listen [to what I say to you]!

Hymn

§2 (A i 6'-20') [You, 0 Sun-goddess of Arinna, are an honored] goddess. [To you, my goddess,] there are revered temples in Hatti, but in no other land are there [any] such for you. [Only in Hatti they] provide for [pure and holy] festivals and rituals for you, [but in no other land] do they provide any such [for you. Lofty temples adorned] with silver and gold [you have only in Hatti, and in no other land] is there anything for you. [Cups and rhyta of silver,] gold, and precious stones you have only in Hatti.

P. Norrie, *A History of Disease in Ancient Times*,
DOI 10.1007/978-3-319-28937-3

Only in Hatti they celebrate festivals for you—the festival of the month], festivals throughout the course of the year, [autumn, winter] and spring, and the festivals of the sacrificial rituals. In no other land do they perform anything for you.

§3 (A i 21'-28') Your divinity, 0 Sun-goddess of Arinna, is honored only in Hatti. Only in Hatti is Mursili, the king, your servant, respectful to you. They perform fully substitute rites, rituals, and festivals for you, 0 Sun-goddess of Arinna. Everything they present to you is pure. Furthermore, the silver and gold in your temples is treated with reverence, and no one approaches it.

§4 (A i 29'- ii 2') You, 0 Sun-goddess of Arinna, are an honored goddess. Your name is honored among names, and your divinity is honored among gods. Furthermore, among the gods you are the most honored and the greatest. There is no other god more honored or greater than you. You are the lord (*sic*) of just judgment. You control the kingship of heaven and earth. You set the borders of the lands. You listen to prayers. You, 0 Sun-goddess of Arinna, are a merciful goddess and you have pity. The divinely guided person is dear to you, 0 Sun-goddess of Arinna, and you, 0 Sun-goddess of Arinna, exalt him. Within the circumference of heaven and earth you, 0 Sun-goddess of Arinna, are the source of light. Throughout the lands you are a favored deity, and you are father and mother to all the lands. You arc the divinely guided lord (*sic*) of judgment, and in the place of judgment there is no tiring of you. Also among the primeval gods you are favored. You, 0 Sun-goddess of Arinna, allot the sacrifices to the gods, and the share of the primeval gods you allot as well. They open up the door of heaven for you, and you cross the gate of heaven, 0 favored [Sun-goddess of Arinna]. The gods of heaven [and earth bow down to you], 0 Sun-goddess of Arinna. Whatever you say, 0 Sun-goddess of Arinna, [the gods] fall down before you, 0 Sun-goddess of Arinna.

Few lines missing to the end of col. i. The beginning of col. ii is completed by 544/u (Güterbock 1980). The line numeration of Lebrun 1980 is indicated in brackets.

§5 (A ii 1–9 [1'-2']) The person at whom the gods are angry and whom they reject, you, 0 Sun-goddess of Arinna, have pity on him! And now, sustain Mursili, the king, [your servant], and [take] Mursili, the king, your servant, by the hand! And to [the words] which Mursili, the king, keeps telling you, hold [your ear] and listen to them!

Pleading

§6 (A ii 10–17 [3'-10') 0 gods, What is this that you have done? You have allowed a plague into Hatti, and the whole of Hatti is dying. No one prepares for you the offering bread and the libation anymore. The plowmen who used to work the fallow fields of the gods have died, so they do not work or reap the fields of the gods. The grinding women who used to make the offering bread for the gods have died, so they do not [make] the god's offering bread any longer.

§7 (A ii 18–44 [11'-37']) The cowherds and shepherds of the corrals and sheepfolds from which they used to select sacrificial cattle and sheep are dead, so that the corrals and sheepfolds are neglected. So it has come to pass that the offering bread, the libations, and the offering of animals have stopped. And you, 0 gods, proceed to hold the sin against us in that matter. To mankind, our] wisdom has been lost, and whatever we do right comes to nothing. 0 gods, whatever sin you perceive, either let a man of god come [and declare it], or let the old women, [the diviners, or the augurs establish it], or let ordinary persons see it in a dream. We shall stroke(?) by means of the thorns(?)/pins(?) of a sarpa. 0 gods, [again] have pity on the land of Hatti. On the one hand it is oppressed with the plague, [and on the other] it is oppressed by hostility. The protectorates which are round about, Mittanni and [Arzawa], are all in conflict, and they do not respect [the gods]. They have transgressed the oath of the gods, and they wish to despoil the temples of the gods. May this become an additional (reason) for the gods' vengeance. Turn the plague, the hostility, the famine, and the severe fever towards Mittanni and Arzawa. Rested are the belligerent lands, but Hatti is a weary land. Unhitch the weary one, and hitch up the rested one.

§8 (A ii 45–55 [38'-48']) Moreover, those lands which belong to Hatti, the Kaska land—they were swineherds and weavers—Arawanna, Kalasma, Lukka, and Pitassa, have declared themselves free from the Sun-goddess of Arinna. They discontinue (the payment of) their tributes and began to attack Hatti. In the past, Hatti, with the help of the Sun-goddess of Arinna, used to maul the surrounding lands like a lion. Moreover, Aleppo and Babylon which they destroyed, they took their goods—silver, gold, and gods—of all the lands, and they deposited it before the Sun-goddess of Arinna.

§9 (A ii 56–60 [49'-53']) But now, all the surrounding lands have begun to attack Hatti. Let this become a further reason for vengeance for the Sun-goddess of Arinna. Goddess, do not degrade your own name!

§10 (A ii 61–67 [54'-60']) Whoever is a cause of rage and anger to the gods, and whoever is not respectful to the gods, let not the good ones perish with the evil ones. Whether it is a single town, a single house, or a single person, 0 gods, destroy only that one! [Look upon] Hatti [with pity, and give the evil plague to other lands.]

Some ten lines to the end of col. ii are missing. They may be completed from the parallel prayer to Telipinu (no. 9, §§ 10–13).

§11' (A iii 1–44) [Some] wish [to burn down your temples]; others wish to take away your rhyta, [cups], and objects of [silver and gold]; others wish to lay waste your fields, your gardens, and your groves; others wish to capture your plowmen, gardeners, and grinding-women. To those enemy lands give severe fever, plague, and famine, 0 Sun-goddess of Arinna, my lady! And you yourself, 0 Sun-goddess of Arinna, let yourself be invoked! [... let] the oppressed become fit [again]. To Mursili, the king, and to the land of Hatti turn [with favour]! Grant to Mursili [and to the land of Hatti] life, health. [vigor, brightness of] spirit forever, and longevity!

Five destroyed lines which may be completed from the parallel prayer, no. 9, § 14.

Grant forever growth of grain, [vines, fruit-trees(?), cattle], sheep, horses [...].

Six destroyed lines which may partly be restored from no. 9, §11:

[Give them a man's valiant,] battle-ready, divine weapon! Put beneath their feet the enemy lands, and [may they destroy them].

0 Sun-goddess of Arinna, [have] pity on Hatti. [...]. [...] winds [...]. May the winds of prosperity come, [and may the land of Hatti grow and] prosper. And to you, 0 gods, your offering bread and your libations will be presented. And the congregation cries out: "[So be it]!"

Colophon

(A iv 1'-8') *Few lines missing* ...] to invoke [the Sun-goddess of Arinna ...] I then recorded the words of the tablet. I have invoked the Sun-goddess of Arinna in Hattusa for seven days, and I have also invoked her for seven days in Arinna, and I spoke these words. There is in addition a sep-arate tablet of the invocation.

(E iv 2'-7') [...] "When [the people] of Hatti [...] are dying [... "; Copy [tablet]: "When they invoke the Sun-goddess of Arinna [...], they speak [these words]."

No. 9 Mursili's Hymn and Prayer to the God Telipinu (CTH 377)

Invocation
§1 (i 1–2) [This] tablet the scribe shall read out daily to the god and shall praise the god saying: §2 (i 3–7) 0 Telipinu, a mighty and honored god are you! Mursili the king, your servant, sent me and your maid-servant the queen, they sent me, saying: "Go, invoke Telipinu, our lord, our personal god (lit. of our head) saying:"

§3 (i 8–10) Whether you, 0 honored Telipinu, are above in heaven among the gods, or in the sea, or gone to the mountains to roam, or if you have gone to an enemy land for battle,

§4 (i 11–17) now let the sweet odor, the cedar and the oil summon you. Return to your temple! I am herewith invoking you by means of offering bread and libation. So be pacified and let your ear be turned to what I say to you, 0 god, and listen to it!

Hymn
§5 (i 18–24) You, Telipinu, are an honored god. To you, my god, there are revered temples only in Hatti, but in no other land are there any such for you. Only in Hatti they provide for pure and holy festivals and rituals for you, but in no other land do they provide any such for you.

§6 (i 25-ii 2) Lofty temples adorned with silver and gold you have only in Hatti, and in no other land are there any such for you. [Cups] and rhyta of silver, gold, and precious stones you have only in Hatti.

§7 (ii 3–8) Only in Hatti they celebrate(!) festivals for you—the festival of the month, festivals throughout the course of the year, winter, spring and fall, and the festivals of the sacrificial rituals. In no other land do they perform anything for you.

§8 (ii 9–19) Your divinity, Telipinu, is honored [only in Hatti]. It is in the land of Hatti that Mursili, the king, your servant, the queen, your maid-servant, and the princes, your servants, are respectful to you. They perform fully your substitute rites, rituals, and festivals for you, 0 Telipinu. Everything they present to you is holy and pure. Furthermore, your rhyta, your cups and your objects in your temples are treated with reverence. [They are] counted over and no one approaches the objects.

§9 (ii 20–22) [You,] Telipinu, are an honored god. [Your] name is honored among names, [and your divinity] is honored among gods.

The rest of col. ii is broken off It was probably similar, though shorter, than the parallel passage in the prayer to the Sun-goddess of Arinna (no. 8, §4-5).

Pleading

§10 (iii 2'-8') [... Turn] with benevolence toward [...]. 0 Telipinu, mighty god, keep alive the king, the queen and the princes, and give them life forever, health, longevity and vigor! [Give] them in their soul [gentleness(?)], radiance and joy!

§ 11 (iii 9'-15') Give them sons and daughters, grandchildren and great-grandchildren! Give them contentment(?) and obedience(?). Give them the growth of grain, vines, cattle, sheep and mankind. Give them a man's valiant, battle-ready, divine weapon! Put beneath their feet the enemy lands, and [may they destroy them].

§12 (iii 6–17) But from Hatti [drive out] the evil fever, plague, famine, and locusts.

§13 (iii 18-iv 8) The enemy lands which are quarrelling and at odds, some are not respectful to you, 0 Telipinu, or to the gods of Hatti; others wish to burn down your temples; others wish to take away your rhyta, cups, and objects of silver and gold; others wish to by waste your fallow lands, vineyards, gardens and groves; others wish to capture your plowmen, vinedressers. gardeners and grinding-women. To those enemy lands give severe fever, plague, famine and locusts.

§14 (iv 9–18) Grant to the king, the queen, the princes and the land of Haiti life, health, vigor, longevity, and brightness of spirit forever! Grant forever growth of grain, vines, fruit-trees(?), cattle, sheep, goats, pigs, mules, asses (var.: horses), together with the beasts of the field, and mankind. May they grow! The rains [...]. May the winds of prosperity come, and in the land of Hatti may everything grow and prosper! And the congregation cries out: "So be it!"

Colophon

(iv 19–21) One tablet. Complete. When the scribe presents daily a plea on behalf of the king before Telipinu.

No. 10 Mursili's "Third" plague Prayer to the Sun-goddess of Arinna (CTH 378.111)

§1 (obv. 1–6) 0 Sun-goddess of Arinna, my lady! 0 gods, my lords! What is this [you have done]? You have allowed a plague into Hatti, so that Hatti has been badly oppressed [by the plague. People kept dying] at the time of my father, at the time of my brother, and now since I have become priest of the gods, they keep on dying [in my time]. For twenty years now people have been dying [in great numbers] in Hatti. Hatti [has been very badly damaged] by the plague.

§2 (obv. 7–13) Hatti has been very much oppressed by the plague. [If someone] produces a child, [the ...] of the plague [snatches (?)] it from him. Should he reach adulthood, he will not attain old age. [And even if old age(?)] will be left for someone, he [will be oppressed(?) by] the plague. He will not [return] to his previous condition. When he reaches old age, [he will ...], but he will not keep warm.

The rest of the obverse and a large portion of the reverse are lost.

§3' (rev. 2'-14') I, Mursili. [your priest, your servant,] hereby plead my case. Hear] me 0 gods, my lords! [Send away] the worry from my heart, [take away the anguish from my soul?] Let the plague [be removed] from Hatti, and send it to the enemy lands. In Hatti [...]. But if the gods, my lords, [do not remove] the plague [from Hatti], the makers of offering bread and the libation pourers will keep on dying. And if they too die, [the offering bread] and the libation will be cut off from the gods, my lords. Then you, 0 gods, [my lords], will proceed to hold the sin against me, saying: "Why [don't you give us] offering bread and libation?" May the gods, my lords, again have pity on Hatti, and send the plague away. [May the plague subside] in Hatti. May it thrive and grow and [return to] its previous condition.

No. 11 Mursili's "Second" Plague Prayer to the Storm-god of Hatti (CTH 378.11)

§1 (C i 1–18) 0 Storm-god of Hatti, my lord! [0 gods], my lords! Mursili, your servant, has sent me saying: "Go speak to the Storm-god of Hatti, my lord, and to the gods, my lords": What is this that you have done? You have allowed a plague into Hatti, so that Hatti has been very badly oppressed by the plague. People kept dying in the time of my father, in the time of my brother, and since I have become priest of the gods, they keep on dying in my time. For twenty years now people have been dying in Hatti. Will the plague never be removed from Hatti? I cannot control the worry of my heart, I can no longer control the anguish of my soul.

§2 (C i 19–28; A obv. 1'-5') When I celebrated the festivals, I busied myself for all the gods. I did not pick out any single temple. I have repeatedly pled to all the gods concerning the plague, and I have repeatedly made vows [to them] saying: "Listen [to me 0 gods], my [lords, and send away] the plague from Hatti. Hatti can [no longer bear this plague. Let the matter on account of which] it has been decimated [either be established through an oracle], or [let me see] it [in a dream, or let a man of god] declare [it]."But the gods [did not listen] to me, [and] the plague has not subsided in Hatti. [Hatti has been severely oppressed by the plague].

§3 (A obv. 6'-12') [The few] makers of offering bread [and libation pourers] of the gods who still remained died off. [The matter of the plague] continued to trouble [me, and I inquired about it] to the god [through an oracle]. [I found] two old tablets: one tablet dealt with [the ritual of the Mala River]. Earlier kings performed the ritual of the Mala River, but because [people have been dying] in Hatti since the days of my father, we never performed [the ritual] of the Mala River.

§4 (obv. 13'-24') The second tablet dealt with the town of Kurustamma: how the Storm-god of Hatti carried the men of Kurustamma to Egyptian territory and how the Storm-god of Hatti made a treaty between them and the men of Hatti, so that they were put under oath by the Storm-god of Hatti. Since the men of Hatti and the men of Egypt were bound by the oath of the Storm-god of Hatti, and the men of Hatti proceeded to get the upper hand, the men of Hatti thereby suddenly transgressed the oath of the gods. My father sent infantry and chariotry, and they attacked the borderland of Egypt, the land of Amqa. And again he sent, and again they attacked. When the men of Egypt became afraid, they came and asked my father outright for his son for kingship. But when my father gave them his son, as they led him off, they murdered him. My father was appalled and he went to Egyptian territory, attacked the Egyptians, and destroyed the Egyptian infantry and chariotry.

§5 (obv. 25'-34') At that time too the Storm-god of Hatti, my lord, by his verdict caused my father to prevail, and he defeated the infantry and the chariotry of Egypt and beat them. But when the prisoners of war who had been captured were led back to Hatti, a plague broke out among the prisoners of war, and [they began] to die. When the prisoners of war were carried off to Hatti, the prisoners of war brought the plague into Hatti. From that day on people have been dying in Hatti. When I found the aforementioned tablet dealing with Egypt, I inquired about it to the god through an oracle saying: "Has this matter been brought about by the Storm-god of Hatti because the men of Egypt and the men of Hatti had been put under oath by the Storm-god of Hatti?"

§6 (A obv. 35'-46'—C iii 3'-7') "And because the *damnassara*-deities were in the temple of the Storm-god, my lord, whereupon the men of Hatti themselves suddenly transgressed the word (of the oath), did this become the cause for the anger of the Storm-god of Hatti, my lord?" And it was confirmed by the oracle. Because of the plague I also asked the oracle about the ritual of the [Mala] River. And then too it was confirmed that I should appear before the Storm-god of Hatti, my lord. I have [just] confessed [the sin before the Storm-god of Hatti]. It is so. We have done [it. But the sin did not] take place in my time. [It took place] in the time of my father [...]. [... that] I know for certain [...]. [...] the matter. [But since] the Storm-god [of Hatti, my lord], is angry about [that matter, and] since people are dying in Hatti, [...] I will keep making [a plea] about it [to] the Storm-god of Hatti, my lord. I kneel down to you and cry for mercy. Hear me, 0 Storm-god of Haiti, my lord! May the plague be removed from Hatti.

§7 (C iii 8'-19'-B iii 16'-24') I will keep removing the causes of the plague which have been established through oracle, and I will keep making restitution for them. With regard to the problem of the oath of the gods which was established as a cause for the plague, I have offered the ritual of the oath for the Storm-god of Hatti, [my lord]. I have also offered [to the gods, my lords]. [I have offered ...] to you, Storm-god of Hatti [...], a ritual for you, [0 gods ...]. As for the [ritual] of the Mala River, which was established for me as a cause for the plague, since I am herewith on my way [to] the Mala River, forgive me, 0 Storm-god of Hatti, my lord, and 0 gods, my lords, for (neglecting) the ritual of the Mala River. I am going to perform the ritual of the Mala River, and I will carry it out. And as for the reason for which I am performing it, namely, because of the plague, have pity on me, 0 gods, my lords, and may the plague subside in Hatti.

§8 (A rev. 10'-19') 0 Storm-god of Hatti, my lord! 0 gods, my lords! So it happens that people always sin. My father sinned as well and he transgressed the word of the Storm-god of Hatti, my lord. But I did not sin in any way. Nevertheless, it so happens that the father's sin comes upon his son, and so the sin of my father came upon me too. I have just confessed it to the Storm-god of Hatti, my lord, and to the gods, my lords. It is so. We have done it. But because I have confessed the sin of my father, may the soul of the Storm-god of Hatti, my lord, and of the gods, ray lords, be appeased again. May you again have pity on me, and send the plague away from Hatti. Let those few makers of offering bread and libation pourers who still remain not die on me.

§9 (rev. 20'-36') I am now continuing to make my plea to the Storm-god, my lord, concerning the plague. Hear me, 0 Storm-god, my lord, and save my life! [I say] to you [as follows]: The bird takes refuge in the cage, and the cage preserves its life. Or if something bothers some servant and he makes a plea to his lord, his lord listens to him, [has pity] on him, and he sets right what was bothering him. Or if some servant has committed a sin, but he confesses the sin before his lord, his lord may do with him whatever he wishes: but since he has confessed his sin before his lord, his lord's soul is appeased, and the lord will not call that servant to account. I have confessed the sin of my father. It is so. I have done it. If there is some restitution (to be made), then there has already [been paid (?)] much for this plague [caused by (?)] the prisoners of war who were brought back from Egyptian territory and by the civilian captives who were brought back. [And] since Hatti has made restitution through the plague, it [has made restitution] for it twenty-fold. Indeed, it has already become that much. And yet the soul of the Storm-god of Hatti, my lord, and of all the gods, my lords, is not at all appeased. Or if you want to require from me some additional restitution, specify it to me in a dream, and I shall give it to you.

§10 (rev. 37'-40') I am now continuing to plead to the Storm-god of Hatti, my lord. Save my life! [And if] perhaps people have been dying for this reason, then during the time that I set it right, let there be no more deaths among those makers of offering bread and libation pourers to the gods who are still left.

§11 (A rev. 41'-44'-C iv 14'-22') [Or] if people have been dying because of some other reason, then let me either see it in a dream, or let it be established through an oracle, or let a man of god declare it, or, according to what I instructed all the priests, they shall regularly sleep holy. 0 Storm-god of Hatti, save my life! Let the gods, my lords, show me their divine power! Let someone see it in a dream. Let the reason for which people have been dying be discovered. We shall stroke(?) by means of the pins(?) of a *sarpa*. 0 Storm-god of Hatti, my lord, save my life, and may the plague be removed from Hatti.

Colophon

(C iv 23–25) One tablet, complete. [How] Mursili made [a plea] because of the plague […].

No. 12 Mursili's "First" Plague Prayer to the Assembly of Gods and Goddesses (CTH 378.1)

§1 (obv. 1–7) [All] you male [gods], all female gods [of heaven(?)], all male gods [of the oath], all female gods of the oath, [all] male primeval [gods], all female (primeval) gods, you gods who have been summoned to assembly for bearing witness to the oath on this [matter], mountains, rivers, springs, and underground watercourses. I, Mursili, [great king(?)], your priest, your servant, herewith plead with you. [Listen] to me 0 gods, my lords, in the matter in which I am making a plea to you!

§2 (obv. 8–15) 0 gods, [my] lords! A plague broke out in Hatti, and Hatti has been severely damaged by the plague. And since for twenty years now in Hatti people have been dying, the affair of Tudhaliya the Younger, son of Tudhaliya, started to weigh on [me]. I inquired about it to the god through an oracle, and the affair of Tudhaliya was confirmed by the deity. Since Tudhaliya the Younger was their lord in Hatti, the princes, the noblemen, the commanders of the thousands, the officers, [the corporals(?)] of Hatti and all [the infantry] and chariotry of Hatti swore an oath to him. My father also swore an oath to him.

§3 (obv. 16–22) [But when my father] wronged Tudhaliya, all [the princes, the noblemen], the commanders of the thousands, and the officers of Hatti [went over] to my father. The deities by whom the oath was sworn [seized] Tudhaliya and they killed [Tudhaliya]. Furthermore, they killed those of his brothers [who stood by] him. [… they sent to Alasiya (Cyprus) and […]. And [since Tudhaliya the Younger] was their [lord], they […] to him […] […] and the lords transgressed the oath […].

§4 (obv. 23–40) [But, you, 0 gods], my [lords], protected my father. […]. And because Hatti [was attacked(?)] by the [enemy, and the enemy] had taken [borderlands] of Hatti, [my father kept attacking the enemy lands] and kept defeating them. He took back the borderlands of Hatti, which [the enemy had taken] and [resettled] them. Furthermore, [he conquered] still other foreign lands [during his] kingship. He sustained Hatti and [secured] its borders on each side. During his reign the entire land of Hatti did well. [Men(?)], cattle and sheep became numerous in his days, and the civilian prisoners who [were brought] from the land of the enemy survived as well. Nothing perished. But now you, 0 gods, [my lords], have eventually taken vengeance on my father for this affair of Tudhaliya the Younger. My father [died(?)] because of the blood of Tudhaliya, and the princes, the noblemen, the commanders of the thousands, and the officers who went over [to my father], they also died because of [that] affair. This same affair also came upon the land of Hatti, and the population of the land of [Hatti] began to perish because of [this] affair. Until now Hatti […], but now the plague [has become] even [worse]. Hatti has been [severely] damaged by the plague, and it has been decimated. I, Mursili, [your servant], cannot [overcome] the worry [of my heart], I can no longer [overcome] the anguish of my soul.

§5 (obv. 41–47) *Very fragmentary passage in which Mursili apparently con-tinues to plead with the oath-deities concerning their vengeance of Tudhaliya's blood. About five more lines, which open a new paragraph, are missing from the end of the obverse. The first seven lines of the reverse, which may belong to the same paragraph, are also very fragmentary.*

§6 (rev. 8'-12') [Now,] I have confessed [it to you, 0 gods(?). Because] my father [killed (?)] Tudhaliya [and ...], my father therefore [performed] a ritual (for the expiation) of blood. But [the land of] Hatti did not [perform.] anything for itself. I performed [the ritual of the blood], but the land did not perform anything. They did nothing on behalf [of] the land.

§7 (rev. 13'-20') Now, because Hatti has been severely oppressed by the plague, and the population of Hatti continues to die, the affair of Tudhaliya has troubled the land. It has been confirmed for me by [the god], and I have further investigated [it] by oracle. They are performing before you, [0 gods], my lords, the ritual of the oath which was confirmed for you, [0 gods], my lords, and for your temples, with regard to the plague of the land and they are clearing [it (i.e. the oath obligation) before] you. And I am making restitution to you, 0 gods, my lords, with reparation and a propitiatory gift on behalf of the land.

§8 (rev. 21'-40') Because you, 0 gods, my lords, [have] taken vengeance for the blood of Tudhaliya, those who killed Tudhaliya [have made] restitution for the blood. But this bloodshed is finished in Hatti again: Hatti too has already made restitution for it. Since it has now come upon me as well, I will also make restitution for it from my household, with restitution and a propitiatory gift. So may the soul of the gods, my lords, again be appeased. May the gods, my lords, again be well disposed toward me, and let me elicit your pity. May you listen to me, to what I plead before you. I have [not] done any evil. Of those who sinned and did the evil, no one of that day is still here. They have already died off. But because the affair of my father has come upon me, I am giving you, 0 gods, my lords, a propitiatory gift on account of the plague of the land, and I am making restitution. I am making restitution to you with a propitiatory gift and reparation. May you gods, my lord, again [have] mercy on me, and let me elicit your pity. Because Hatti has been oppressed by the plague, it has been reduced in size. [And those makers of offering bread and libation pourers who used to prepare] the offering bread and the libation for the gods, my lords, [since Hatti] has been severely oppressed by [the plague], [they have died] from the plague. [The plague] does not subside at all, and they continue to die, [even those] few [makers of offering bread] and libation pourers [who still remain will die, and nobody will prepare] for you offering bread and libation any longer.

§9 (rev. 41'-51') May [you gods, my lords], have mercy on [me again] because of the offering bread and the libation which [they prepare for you], and let me elicit your pity. Send the plague [away from Hatti]. Let those few makers of offering bread [and pourers of libation] who [still remain] with you not he harmed, and let them not go on dying, Let them prepare [the offering bread] and the libation for you. 0 gods, my lords, turn the plague [away, and send] whatever is evil to the enemy land. Whatever has happened in Hatti because of Tudhaliya, send it [away] 0 gods, [my lords]. Send [it] to the enemy land. May you again have mercy on Hatti, and let [the plague] subside. Furthermore, [because] I, your priest, your servant, elicit your pity, may you have mercy on me. Send away the worry from my heart, take away the anguish from my soul!

Colophon

(rev. 52'-53') [One tablet], complete. When Mursili made a plea [because of the plague ...].

No. 13 Mursili's "Fourth" Plague Prayer to the Assembly of Gods (Arranged by Localities) (CTLH 378.IV)

§1 (i I-16) 0 gods, my lords: Noble Storm-god, the two lords of Landa, Iyarri, gods of Hatti, gods of Arinna, gods of Zippalanda, gods of Tuwanuwa, gods of Hupisna, gods of Durmitta, gods of Ankuwa, gods of Samuha, gods of Sarissa, gods of Hurma, gods of Hanhana, gods of Karahna, gods of Illaya, Kamrusepa of Taniwanda, gods of Zarruwisa, Storm-god of Lihzina, Protective-god of the Army Camp of His Majesty's father which is in Marassantiya, Uliliyassi of Parmanna, gods of Kattila, Storm-god of Hasuna, gods of Muwani, gods of Zazzisa, the Telipinu-gods [whose] temples in the land have been destroyed, gods of Salpa, Storm-god of Ar[ziya (?)].

§2 (i 17–20) 0 gods, my lords! I, Mursili, [your servant], your priest, herewith bow down to you. Lend me your ear and hear me in the matter in which I have bowed down to you.

§3 (i 21–35) 0 gods, my lords! Since ages past you have been inclined towards [men] and have [not] abandoned mankind. And mankind [became] populous and your divine servants [were] numerous. They always set up for the gods, [my] lords, offering bread and libation. 0 gods, my lords, you have turned your back on mankind. All of a sudden, in the time of my grandfather Hatti was oppressed, [and it] became [devastated] by the enemy. Mankind was [reduced in number] by plague, and your [servants] were reduced in number. And among you, [gods], my lords, [one had no] temple, and [the temple] of another [fell into ruin]. Whoever [served] before a god perished, and [your] rites [were neglected]. [No] one performed [them] for you.

§4 (i 36–46) [But] when my [father] became king, [you], 0 gods, my lords, stood behind him. He resettled the [depopulated] lands. [And for you], 0 gods, my lords, in whatever temple there were no [objects], or whatever image of god had been destroyed, my father restored what he could, though what he could not, he did not restore. 0 gods, my lords, you never before oppressed my father, and you never before oppressed me. But now you have oppressed me.

§5 (i 47–55) When my father went to Egyptian territory; since that day of Egypt, death has persisted in [Hatti], and from that time Hatti has been dying. My father repeatedly inquired through the oracles, but he did not find you, 0 gods, my lords, through the oracles. I have also repeatedly inquired of you through oracle, but I have not found you, 0 gods, my lords, through oracle.

§6 *In this section the scribe (of manuscript A) left an empty space of about six lines, indicating through the single word "destroyed" that the corresponding passage in the manuscript from which he was copying was damaged.*

§7 (ii 1–3) *Only a few words are preserved from this paragraph (in both copies). It probably dealt with the rites (hazziwita) that Mursili intended to restore.*

§8 (B ii 3'-16') *The first three lines are very fragmentary.* For whatever [god] there is [a temple], but he has no [objects], I will restore [them for him]. And for whatever god [there is no temple], I will build a temple for him. And whichever [gods] have been destroyed, I will restore for them a statue [… and] its […] as before. *The rest of col. ii and all of col. iii, except the beginnings of lines, are lost.*

§9' (A iv 1–5) Or should I have restored it for [the gods], my lords, from my land, or from my infantry and chariotry? If I should indeed reestablish the gods, since now the members of my household, land, infantry and chariotry keep dying, by what means should I reestablish you, 0 gods?

§10' *As in §6, the scribe (of manuscript A) left an empty space of about ten lines, indicating that the corresponding passage in the 'manuscript from which he was copying was damaged.*

§11' (iv 16–28) And it dies, by what means should I reestablish [you]? 0 gods, have mercy on me again because of this [reason]! Turn(!) towards me! Send the plague away from the land! Let it subside in the towns where people are dying, and let the plague not return to the towns in which it has subsided! I have [said] to myself thus: "If the aforementioned word of the god is true, [and] my father [could not discover them] through an oracle, nor could I discover them [through an oracle], should the land of Hatti [inquire by oracle] and [will it discover] them through an oracle?" And I have pled my case. [...] *The remaining fifteen lines or so are almost entirely lost.*

No. 14 Mursili's "Fifth" Plague Prayer to the Assembly of Gods (Arranged Typologically) (CTH 379)

§1 (i! 1–4) [Sun-god of Heaven], Storm-god [of ... , Sun-goddess of] Arinna, Mezzulla, [Hulla(?)/Zintuhi(?)], Storm-god of Hatti, [Storm-god of] Zippalanta;

§2 (i 5–6) [...], Seri, Hurri, [Storm-god *pihaimi*(?)], all the Storm-gods;

§3 (i 7–8) [...]. Hebat of Kummanni, all [the Hebats], Halki;

§4 (i 9–10) All [the Sarrumas(?)], [...], all the Hebat-Sarrumas;

§5 (i 11–15) Protective-god (LAMMA), [Protective-god of] Hatti, all the Protective-gods, Ishtar, [Ishtar of the Field of] His Majesty, Ishtar of Samuha, [all the] Ishtars, Telipinu, all the Telipinus, War-god (ZABABA), all the War-gods;

§6 (i 16–22) Sun-goddess of the Netherworld, Lelwani, Pirwa, Marduk, Iyarri, Hasammeli, Fate-goddesses, Mother-goddesses, all the male gods of the assembly(!), all the female gods of the assembly(!). the place of assembly, the place in which the gods assemble for judgment.

The rest of the column, about thirty lines, is almost entirely lost. The verbal endings at the end of lines 6"-8" in col. i(!) probably belong to second person plural imperatives, which may be addressed to the "male gods (and) female gods" mentioned in 1. 5."

§7' (ii! 6') [...] [... the tablet of/about] Egypt.

§8' (ii 7'-17') To this tablet I did not add any word, nor did I remove [any]. 0 gods, my lords, take notice? I do not know whether any of those who were kings before me added [any word] to it or removed any. I do not know anything, and I have not heard a word of it since.

§9' (ii 18'-24') I did not concern myself with those borders which were set for us by the Storm-god. Those borders that my father left me, those borders [I kept]. I did [not] desire from him [anything]. Neither [did I take anything] from his borderland.

§10' (ii 25') […] this matter […]. *Gap of about two lines between KBo 31.121 and KBo 31.121a, followed by three fragmentary lines.*

§11' (KUB 31.121a ii 6"-9") [… infantry and] chariotry of Hatti […]. […] He (i.e. Suppiluliuma) sent out Lupakki and Tarhunta-zalma, and they attacked those Lands.

§12' (ii 10"-15") The king of Egypt died in those very [days]. I was still a child, so I did not know whether the king of Egypt lodged [a protest(?)] to my father about those lands, or whether he [did] nothing.

§13' (ii 16"-20") And since the wife of the king of Egypt was a widow, she wrote to my father. […] to talk with women […]. I, in those […] I was not seen(?) […]. *Some eight lines missing at the end of col. ii. All of the reverse is broken off. From the colophon on the edge of KUB 48.111 only "not complete" is preserved.*

About the Author

Dr. Philip Norrie MBBS, MA, MSc, MSocSc[Hons], PhD, MD

Conjoint Senior Lecturer, Faculty of Medicine, University of New South Wales, Sydney

Vice Chairman, Medical Advisory Committee, Northern Cancer Institute, Sydney

Faculty Affiliate, School of Public and Community Health Sciences, College of Health Professions and Biomedical Sciences, University of Montana, USA

Dr. Philip Norrie is a family physician and medical historian from Palm Beach in Sydney, Australia, who believes that disease has not been given its due recognition as a major contributing factor to the history of mankind, a situation he wishes to correct.

Born in 1953 and educated at Knox Grammar School, where he was awarded the History and Geography Prizes each year plus the Sports and Studies Prize for being the best all-rounder in his final year in 1970, Dr. Norrie gained his medical degree [MBBS—Bachelor of Medicine and Bachelor of Surgery] from the University of New South Wales in 1977. This was followed by being the first graduate from the newly formed History and Philosophy of Science Department at Australia's oldest university—the University of Sydney, in 1993 with a thesis on Australian wine history [Macleay family].

In 1998 Dr. Norrie received his Master of Social Science degree with honours with a thesis on Australian wine history [Leo Buring], from

P. Norrie, *A History of Disease in Ancient Times*,
DOI 10.1007/978-3-319-28937-3

Australia's largest wine science school at Charles Sturt University. His PhD followed in 2006 from the University of Western Sydney, with a thesis on the history of wine as a medicine for the past 5,000 years, showing that wine is our oldest medicine, most documented medicine and best preventative medicine.

2007 saw Dr. Norrie gain a Master of Arts degree from the University of Sydney with a thesis analyzing the causes of death in Australia's oldest prison at Darlinghurst Gaol in Sydney and showing that Charles Dickens was wrong—the death rate in gaol during the Victorian era was much better than in the general population.

In 2014 Dr. Norrie gained his Doctor of Medicine degree from the University of New South Wales, with a thesis on the role of disease during the end of the Bronze Age in the Near East and its role in the demise of the Hittite Empire.

Dr. Norrie has also published 17 books about wine and medical history.

After completing his PhD Dr. Norrie decided to take the topic of wine and health one step further by making wine even healthier by inventing the world's first full strength Resveratrol Enhanced Wine [REW—arguably the world's healthiest beverage]. Resveratrol is the anti-oxidant in wine, derived from the grape skin, that makes wine so healthy and is also the only substance to stimulate Sirtuin production, which makes cells live longer. Normal white wine contains 1–2 mg/l of resveratrol and normal red wine contains 3–6 mg/l of resveratrol; while Dr. Norrie's REW contains 100 mg/l in it; thus making it much more therapeutic.

Dr. Norrie's current research is into the role of disease in the demise of the Sumerian and Indus Valley Civilizations—a topic so far ignored by ancient historians and archaeologists.

Dr. Norrie married his physiotherapist wife Belinda whilst still in Medical School. They have two sons—Andrew, who is a barrister and Alexander, who is a lawyer; plus their Borzoi dog Sasha. Dr. Norrie's interests include wine, wine and medical history plus collecting the autographs of famous doctors, medals, toy trains, Meccano, Dinky toys and slot cars—all the fun of his youth!

Website—www.drnorrie.info

Bibliography

(A) Books

(1) *General Ancient History Reference Books*

Bachhuber, C., & Roberts, R. G. (Eds.). (2009). *Forces of transformation – The end of the Bronze Age in the Mediterranean.* Oxford: Oxbow Books.

Barnard, L. (Ed.). (2010). *Historical atlas.* Sydney: Millenium House.

Black, G. (Ed.). (2002). *Encyclopedia of world history.* Bath: Parragon Publishing.

Blainey, G. (2001). *A short history of the world.* Camberwell/Victoria: Viking/Penguin Group (Australia).

Brothwell, D. R., & Chiarelli, B. A. (Eds.). (1973). *Population biology of the ancient Egyptians.* London: Academic Press.

Burenhult, G. (Ed.). (2004). *Great civilizations.* San Francisco: Fog City Press.

Cline, E. H. (Ed.). (2010). *The Oxford handbook of the Bronze Age Aegean (ca. 3000–1000 BC).* Oxford: Oxford University Press. Chapter 36 titled 'The Collapse at the end of the Bronze Age' by Durham University's Oliver Dickinson gave a very useful summary of current thinking about the causes of the collapse including noting the abandonment of sites. Plague is mentioned but the conclusion is that it was most likely multifactorial.

Cockburn, A., & Cockburn, E. (Eds.). (1998). *Mummies, disease and ancient cultures.* Cambridge: Cambridge University Press.

Cunliffe, B. (Ed.). (1994). *The Oxford illustrated prehistory of Europe.* Oxford: Oxford University Press.

Daniels, P. (2003). *Almanac of world history.* Washington, DC: National Geographic.

Danzer, G. (2002). *Atlas of world history.* Rochester: Grange Books.

Desborough, V. R.'. A. (1972). *The Greek Dark Ages.* London: Ernest Benn.

P. Norrie, *A History of Disease in Ancient Times*,
DOI 10.1007/978-3-319-28937-3

Diamond, J. (1997). *Guns germs and steel.* London: Viking for Random House.

Dickinson, O. (2006). *The Aegean from Bronze Age to Iron Age.* Abingdon: Routledge. Chapter 2 titled 'The Collapse of the Bronze Age Civilization' gave a good overview of the subject similar to what he wrote in chapter 36 above.

Dupuy, R. E., & Dupuy, T. N. (1993). *The Harper encyclopedia of military history.* New York: Harper Collins Publishers.

Farrington, K. (2002). *Historical atlas of empires.* London: Mercury Books.

Finlay, M. I. (1981). *Early Greece: The Bronze and Archaic Ages* (New and Rev. ed.). London: W.W.Norton & Co.

Gardiner, A. (2005). *The Egyptians.* London: The Folio Society, in Conjunction with Oxford University Press.

Goetz, P. W. (Ed.). (1989). *The New Encyclopedia Britannica.* Chicago: University of Chicago.

Grant, M. (1969). *The ancient Mediterranean.* New York: Charles Scriber's Sons.

Hendel, R. (2005). *Remembering Abraham—Culture, memory, and history in the Hebrew Bible.* Oxford: Oxford University Press.

Isaacs, A. (2000). *Oxford dictionary of world history.* Oxford: Oxford University Press.

Kozloff, A. P. (1992). *Egypt's dazzling sun: Amenhotep III and his world.* Cleveland: Cleveland Museum of Art. Kozloff provides an in depth look at the reign of Amenhotep III and shows, through examination of artifacts from this period of Egyptian history, that some major destructive force adversely affected his reign. This force was most likely bubonic plague. It is a very thorough work and is used in this book.

Kuhrt, A. (1995). *The ancient Near East c. 3000–330 BC volume one.* London: Routledge.

Lemche, N. P. (1998). *The Israelites in history and tradition.* London: SPCK.

Pernicka, E., et al. (2003). Early Bronze Age metallurgy in Northeast Aegean. In G. A. Wagner et al. (Eds.), *Troia and Troad: Scientific approaches.* London: Springer, Natural Science in Archaeology.

Roberts, J. M. (2002). *Ancient history.* London: Duncan Baird Publishers.

Roberts, J. M. (2003). *The new history of the world.* Oxford: Oxford University Press.

Ruffer, M. A. (1921). *Studies in the paleopathology of Egypt.* Chicago: University of Chicago Press. Ruffer examined many mummies and other Egyptian material, thus founding the science of Paleopathology, in this book. He showed that Egyptian mummies showed evidence of smallpox, tuberculosis, schistosomiasis and polio for example. It is a very thorough work and is a major resource for this book.

Scarre, C. (1993). *Timelines of the ancient world.* London: Dorling Kindersley.

Stiebing, W. H. (2009). *Ancient Near Eastern history and culture.* New York: Pearson Longman. This book provided some material about the Plague of the Philistines and some general background information about the early Israelite nation.

Van De Mieroop, M. (2008). *A history of the ancient Near East ca. 3000–323 BC*. Oxford: Blackwell Publishing.

Wagner, G. A. et al. (Eds). (2003). *Troia and Troad: Scientific approaches*. London: Springer, Natural Science in Archaeology.

Waldbaum, J. C. (1978). *From bronze to iron. Studies in Mediterranean archaeology*. Goteburg: Paul Astroms Forlag.

Yasur-Landau, A. (2010). *The Philistines and Aegean migration at the end of the late Bronze Age*. Cambridge: Cambridge University Press.

(2) *Medical History Books*

Ackerknecht, E. H. (1982). *A short history of medicine*. Baltimore: The Johns Hopkins University Press.

Barnard, B. (2005). *Outbreak—Plagues that changed history*. New York: Crown.

Barnes, E. (2006). *Diseases and human evolution*. Albuquerque: University of New Mexico Press.

Beck, R. W. (2000). *A chronology of microbiology in historical context*. Washington: American Society of Microbiology Press.

Bollet, A. J. (2004). *Plagues and poxes – The impact of human history on epidemic disease*. New York: Demos.

Burke, R. M. (1938). *A historical chronology of tuberculosis*. Springfield: Charles C. Thomas Publishers. This book described tuberculosis in ancient Egypt, hence was used in this book.

Bynum, W. F., & Porter, R. (1993). *Companion encyclopedia of the history of medicine—Volume 1 and 2*. London: Routledge Reference.

Cartwright, F. F., & Biddiss, M. (2004). *Disease and history*. Stroud: Sutton Publishing.

Clendening, L. (1960). *Source book of medical history*. New York: Dover Publications.

Davies, G. (Ed.). (2000). *Timetables of medicine*. New York Black Dog and Leventhal.

Davis, L. (2008). *Natural disasters*. Facts on File Science Library. Checkmark Books.

Desowitz, R. (1997). *Tropical diseases from 50,000BC to 2,500AD*. London: Flamingo.

Dobson, M. (2007). *Disease*. London: Quercus.

Duffin, J. (2006). *History of medicine*. Toronto: University of Toronto Press.

Farnell, J. (1998). *Invisible enemies—Stories of infectious disease*. New York: Farrar Straus Giroux.

Gregg, C. T. (1985). *Plague*. Albuquerque: University of New Mexico Press.

Grmek, M. D. (1989). *Diseases in ancient Greek world*. Baltimore: The Johns Hopkins University Press.

Haeger, K. (2000). *History of surgery*. London: Harold Starke Publishers.

Hare, R. (1954). *Pomp and pestilence—Infectious disease, its origins and conquest*. London: Gollancz Publishers. In this book Hare continued Wynn's possibility that the 1322 BCE Hittite Empire epidemic was due to smallpox, imported into the Hittite Empire by Egyptian prisoners of war. This assumption is challenged by this book.

Hays, J. N. (2003). *The burdens of disease*. New Brunswick: Rutgers University Press.

Hoff, B. (2000). *Mapping epidemics—A historical atlas of disease*. New York: Franklin Watts for Grolier.

Hopkins, D. R. (2002). *The greatest killer—Smallpox in history*. Chicago: The University of Chicago Press. This book confirmed that Ramesses V had smallpox, as the author was allowed to view the mummy of Ramesses V in person, with the permission of Anwar Sadat, the President of Egypt at the time. This book also says that the 1322 BCE Hittite Epidemic was most likely due to smallpox—a view that this book disputes. A very thorough and well-presented history, written by one of the world's authorities on the subject.

Karlen, A. (1995). *Man and microbes—Disease and plague in history and modern times*. New York: Simon and Schuster.

Kennedy, M. T. (2004). *A brief history of disease*. Mission Viejo: Science and Medicine, Asklepiad Press.

King, H. (Ed.). (2005). *Health in antiquity*. Abingdon: Routledge. Chapter 1 titled 'Disease and the Prehistory of the Aegean' by medical Historian Robert Arnott from Oxford University was a very thorough and useful overview of the subject but Arnott does not make the connection that disease may have been a cause of the end of the Bronze Age in the Near East.

Kiple, K. F. (1999). *The Cambridge world of human disease*. Cambridge: Cambridge University Press.

Kiple, K. F. (Ed.). (2003). *The Cambridge historical dictionary of disease* (2003rd ed.). Cambridge: Cambridge University Press.

Kohn, G. C. (Ed.). (2001). *Encyclopedia of plagues and pestilence*. New York: Checkmark Books.

Loudon, I. (Ed.). (1997). *Western medicine*. Oxford: Oxford University Press.

Lyons, A. S., & Petrucelli, R. J. (1979). *Medicine—An illustrated history*. South Melbourne: The Macmillan Company of Australia.

Major, R. H. (1945). *Classic descriptions of disease*. Springfield: Charles C. Thomas Publisher.

McNeill, W. H. (1998). *Plagues and peoples*. New York: Anchor Books for Random House.

Oldstone, M. B. (2000). *Viruses, plagues and history*. Oxford: Oxford University Press.

Porter, R. (1999). *The greatest benefit to mankind—A medical history of humanity*. New York: W.W.Norton and Co.

Porter, R. (Ed.). (2001). *Cambridge illustrated history of medicine*. Cambridge: Cambridge University Press.

Porter, R. (2004). *Blood and guts—A short history of medicine*. New York: W.W.Norton and Co.

Snodgrass, M. E. (2003). *World epidemics*. London: McFarland and Co. This book provided some brief descriptions of early epidemics and diseases around the time period of this book such as polio, smallpox on the face of Ramesses V and the plague of the Philistines.

Watts, S. (1999). *Epidemics and history*. New Haven: Yale University Press.

Watts, S. (2003). *Disease and medicine in world history*. New York: Routledge.

Wills, C. (1996a). *Plagues—Their origin, history and future*. London: Harper Collins.

Wills, C. (1996b). *Yellow fever black goddess—The coevolution of people and plagues*. London: Basic Books.

Zimmerman, B. E., & Zimmerman, D. J. (2003). *Killer germs*. Chicago: Contemporary Books.

(3) *Books About the Hittites and the Sea Peoples*

Brown John, J. J. (1977). *The Hittites—People of a thousand gods*. London: Collins.

Bryce, T. (2004). *Life and society in the Hittite world*. Oxford: Oxford University Press.

Bryce, T. (2008). *Hittite warrior*. Oxford: Osprey Publishing.

Bryce, T. (2010). *The kingdom of the Hittites*. Oxford: Oxford University Press. Trevor Bryce, from the University of Queensland, is the current World expert on the Hittite Empire. This is his latest and most comprehensive book which incorporates all his knowledge and is a good summary of the latest information and current state of knowledge about the Hittite Empire. This he does very well. He devotes a few pages to the 1322 BCE Hittite Empire epidemic, but does not say it was due to bubonic plague as this book advocates. He incorporates the latest theory about Hattusa suggesting that it was simply abandoned over several months, but does not say why, let alone state that it was due to disease. This book was used extensively for latest background information about the Hittites.

Canby, J. V. et al. (Eds). (1986). *Ancient Anatolia—Aspects of change and cultural development* (Essays in Honor of Machteld J. Mellink). The University of Wisconsin Press.

Carruba, O. (1995). *Atti Del II Congresso Internazionale Di Hittitologia*. Pavia: Gianni Iuculano Editore.

Ceram, C. W. (2001). *The secret of the Hittites*. London: Phoenix Press.

Cilingiroglu, A. (1994). *The proceedings of the third Anatolian Iron Ages colloquium held at Van, 6–12 August 1990*. Ankara: The British Institute of Archaeology.

Gurney, O. R. (2005). *The Hittites*. London: The Folio Society.

Klengel, H. (1999). *Geschichte des Hethitischen Reiches*. Leiden: Brill. This book provides a German archaeological insight into the world of the Hittites. Klengel mentioned that plagues were often mentioned in late Bronze Age Syrian documents, but did not specify which ones unfortunately.

Lloyd, S. (1989). *Ancient Turkey—A traveler's history of Anatolia*. Berkeley: University of California Press.

Macqueen, J. G. (2001). *The Hittites*. London: Thames and Hudson.

Ozguc, T. *The Hittites*. Donmez Offset, Ankara for The Museum of Anatolian Civilizations.

Peiser, B. J. (1998). *Natural catastrophes during Bronze Age civilizations—Archaeological, geological, astronomical and cultural perspectives.* Oxford: Archaeopress. This book summarized all the papers presented at the conference and was most useful for discussing alternative causes for the end of the Bronze Age such as volcanic activity, comets and earthquakes, but disease was not mentioned at all—hence the need for this book to present yet another possible cause.

Sandars, N. K. (1978). *The sea peoples—Warriors of the ancient Mediterranean 1250–1150 BC.* London: Thames and Hudson.

Van Den Hout, T. P. J. (Ed.). (2006). *The life and times of Hattusili III and Tuthaliya IV.* Leiden: Nederland's Institute Voor Het Nabije Oosten.

Velikovsky, I. (1977). *Peoples of the sea.* London: Sidgwick and Jackson.

Versuch, E. (1984). *Griechische Fruhgeschichte.* Wien: Osterreichischen Akademie Der Wissenschaften.

Ward, W. A. (Ed.). (1992). *The crisis years: The 12th century B.C. – From beyond the Danube to the Tigris.* Dubuque: Kendall Hunt Publishing Company. Even this detailed summary of conference papers did not mention disease at all.

Yakar, J. (2000). *Ethnoarchaeology of Anatolia—Rural socio-economy in the Bronze and Iron Ages.* Tel Aviv: Emery and Claire Yass Publications in Archaeology, Institute of Archaeology, Tel Aviv University.

(4) *Books About the End of the Bronze Age*

Cline, E. H. (2014). *1177 B.C.—The year civilization collapsed.* Princeton/Oxford: Princeton University Press. This book provides a very good summary of current conventional thinking about how the Near Eastern Bronze Age ended, but it does not factor in the potential role of infectious disease unfortunately.

Drews, R. (1993). *The end of the Bronze Age.* Princeton: Princeton University Press. In this book Drews argues that changes in warfare from the large palatial chariot based armies to the new better equipped foot soldier based forces with iron body protective armor and iron swords and spears; meant the end of the palatial empires. Drews does not mention disease at all in his book.

Slattery, T. (2000). *The tragic end of the Bronze Age.* San Jose: Writers Club Press. This is the only work which puts forward the hypothesis that disease contributed to the end of the Bronze Age. Unfortunately Slattery only mentions smallpox and no other diseases, whereas this book will suggest another nine diseases as possible causes for the end of the Bronze Age and Hittite Empire. In the book Slattery uses many Biblical references and tends to wander off the subject a lot.

(B) Articles

(1) *Articles About the Ancient Near East*

History of Ancient Europe, http://www.essentialhumanities.net/his1.php, "*History of Ancient Europe*". Accessed 15 Oct 2011.

Late Bronze Age, http://cuttingedgeminitures.com/Late-Bronze-Age. Accessed 9 May 2012.

Lehmann, G. A. (1985). Vortage, Rheinisch-Westfalische Akademie der Wissenschaften: Geisteswissen-schaften; G276, Westdeutscher Verlag.

The Mycenaean civilizations, http://www.aroundgreece.ccm/ancient-greece-history/mycenaean-civilization-greece.php. Accessed 3 Dec 2013.

(2) *Articles About Ancient Egypt*

Akhenaten, http://www.ancientegyptonline.co.uk/Akhenaten.html. Accessed 3 Dec 2013.

Amenhotep III, http://euler.slu.edu/~bart/egyptianhtml/kings%20and%20Queens/amanhotepiii.htm. Accessed 7 Dec 2012.

Ancient Armana, http://heritage-key.com/egypt/ancient-amarna-black-death-london-ancient-history-plague. Accessed 21 Jan 2012. This article again discussed infectious disease at the time of the Amarna letters, thus supporting the other Amarna letter articles.

Ancient Egyptian medicine, http://www.reshafim.org.il/ad/egypt/timelines/topics/medicine.htm. Accessed 10 Oct 2011.

Amarna, http://www.kchanson/Amarna_tablet_35. Accessed 3 Dec 2013. This article provided more evidence, through Amarna letters, of the presence of infectious disease in various places around the Near East in the time frame of this book.

Armitage, P. L. (1994). Unwelcome companions: Ancient rats reviewed. *Antiquity*, *68*, 231–240. This article examined the remains of rats from ancient Egypt and showed that bubonic plague carrying rats existed then.

Battle of Kadesh, http://www.allaboutturkey.com/kades.htm. This article discussed the first recorded major battle in the ancient world which was the Battle of Kadesh between the armies of the Hittite Empire and ancient Egyptian Empire. The battle occurred where these great power's borders met—in Syria near the town of Kadesh and was inconclusive as to who actually won, as both sides claimed victory. It raises the question that if the Hittite Empire had more soldiers, instead of losing so many in the 1322 BCE Hittite epidemic, could they have had a major victory and then been able to go on and invade Egypt itself and not just occupy some land in one of its far off provinces to the north. Accessed 3 Dec 2013.

BBC-History-Ancient history in depth: The end of the Amarna Period (pp. 1–6). Accessed 3 Dec 2013.

Costantini, L. (1975). Typology and socioeconomical implications of entomological finds from some ancient Near Eastern sites. *Paleorient*, *3*, 247–258.

Egypt's new kingdom and final decline (1778–525 BC), http://www.flowofhistory.com/units/pre/2/fc11b. Accessed 3 Dec 2013.

Egyptian Hittite correspondence, http://www.reshafim.org.il/ad/egypt/egyptian-hittite_correspondence.htm. Accessed 24 Sept 2011.

Egyptian Hittite Peace Treaty, http://www.milestonedocuments.com/documents/view/Egyptian-hittite-peace-treaty. Accessed 3 Dec 2013.

Kozloff, A. P. (2006). Bubonic plague during the Reign of Amenhotep III? *KMT A Modern Journal of Ancient Egypt*, *17*(3), 36–46 and 83–84. This article is an excellent resource for this book and was quoted extensively. In it Kozloff argues, very thoughtfully and thoroughly with great examples, that after examining the quality of the tombs and artifacts from the reign of Amenhotep III it is apparent that some major disaster adversely affected Egypt during the years 12 to 20 of his reign. Later on in the article Kozloff suggests that the adverse event was not a war or invasion by another country or a famine but an infectious disease epidemic—most likely bubonic plague. This article shows that there was a reservoir for the bubonic plague in Egypt which later infected the Hittite Empire as the 1322 BCE Hittite epidemic and later still as a possible cause for the end of the Bronze Age and Hittite Empire.

Malkata Palace, http://www.ancientegyptonline.co.uk/malkatapalace.html. Accessed 3 Dec 2013.

Mummies: Mummies and disease in Egypt, http://www.uic.edu/classes/osci/osci590/6_2Mummies%20Mummies%20and%20Disease%20in%20Egypt.htm this article discussed various diseases seen in ancient Egyptian mummies and thus supports Ruffer's work by showing that smallpox, tuberculosis and polio for example existed in ancient Egypt. Accessed 10 Oct 2011.

Panagiotakopulu, E. (2001a). Fleas from Pharaonic Amarna. *Antiquity*, *75*, 499–500. This article, written by an expert in the new field of Paleoentomology, showed that it is possible to find the remains of fleas in the archaeological sites of Egypt and to be able to identify these fleas as bubonic plague carrying fleas. Thus it is possible to conclude that bubonic plague existed in Egypt at that time, which helps to support Kozloff's ideas.

Panagiotakopulu, E. (2001b) Ramesses II and the tobacco beetle. *Antiquity*, *75*, 549–556.

Panagiotakopulu, E. (2003). Insect remains from the collections in the Egyptian Museum of Turin. *Archaeometry 45*(Part 3), 355–362.

Panagiotakopulu, E. (2004). Pharaonic Egypt and the origins of plague. *Journal of Biogeography*, *31*, 269–275. This article uses Paleoentomology to examine the archaeological evidence to show that bubonic plague may have entered ancient Egypt from trading vessels from India and a lot earlier than had been thought previously. The article also shows that bubonic plague may have started in India and not China as previously believed.

Panagiotakopulu, E. (2009). Environment, insects and the archaeology of Egypt. In S. Ikram & A. Dodson (Eds.), *Beyond the horizon: Studies in Egyptian art, archaeology and history in honor of Barry J. Kemp* (pp. 347–361). Cairo: American University of Cairo Press.

Panagiotakopulu, E. (2010). Underneath Ranefer's floors—Urban environments on the desert edge. *Journal of Archaeological Science*, *37*, 474–481. This article examined the remains of rats and other creatures found under the floor of a house in the workers camp at Amarna. It proved that bubonic plague carrying fleas and rats existed there; again more evidence to support Kozloff.

Ramesses the Great—The Pharaoh who made peace with his enemies, http://www.arabworldbooks.com/rameses.htm. Accessed 24 Sept 2011.

Sed, http://www.digitalegypt.ucl.ac.uk/ideology/sed/index.html. Accessed 3 Dec 2013.

Sekhmet, http://ancientegyptonline.co.uk/Sekhmet.html. Accessed 3 Dec 2013.

Sekhmet, http://www.egyptianmyths.net/sekhmet.htm. Accessed 9 Dec 2012.

Sakhmet, http://www.pantheon.org/articles/s/sakhmet.html. Accessed 3 Dec 2013.

The Pharaoh who made peace with his enemies and the first peace treaty in history, http://www.touregypt.net/featurestories/treaty.htm. Accessed 3 Dec 2013.

The rise and fall of civilizations, http://www.gold-eagle.com/article/rise-and-fall-civilizations-0. Accessed 3 Dec 2013.

Treaty of Kadesh, http://www.istanbularkeoloji.gov.tr/web/30-125-1-1/muze__en/collections/ancient_orient_museum_artifacts/treaty_of_kadesh

(3) *Articles About the Hittites.*

Arda, B. (2009). Anatolia; The cradle of modern medicine. *Ankara Universitesi Tip Falultesi Mecmuasi, 62*(1), 9–10.

Forrer, E. O. (1937). The Hittites in Palestine. *Palestine Exploration Quarterly, 69*, 100–115. Forrer was a prominent early Hittitologist and in this article he examines the then known translated Hittite tablets and mentions the 1322 BCE Hittite epidemic and states that it was brought into the Hittite Empire by Egyptian prisoners of war. One must remember that when this article was written the Hittite tablets had only been discovered a decade or so earlier and only some had been translated, so this article was a very early reference about the 1322 BCE Hittite epidemic.

Hattusa, http://creationwiki.org/Hattusa. Accessed 3 Dec 2013.

Hattusa, http://heritage-key.com/site/hattusa. Accessed 9 Apr 2012.

History of the Hittites, http://www.ancient.eu.com/article/169/. Accessed 3 Dec 2013.

History of the Hittites, http://www.historyworld.net/wrldhis/PlainTextHistories.asp?historyid=ab66. Accessed 3 Dec 2013.

Hittites, http://danwagner.homestead.com/files/hittites.htm. Accessed 3 Dec 2013.

Hittites, http://history-world.org/hittites.htm. Accessed 3 Dec 2013.

Hittites, http://i-cias.com/e.o/hittites.htm. Accessed 3 Dec 2013.

Hittites, http://oxfordbibliographiesonline.com/view/document/obo-9780195393361/obo-9780195393361-0075.html. Accessed 1 Jan 2012.

Hittites, http://timelines.ws/countries/hittites.html. Accessed 3 Dec 2013.

Hittites, http://www.hittites.info/history.aspx?text=history%2FEarly+Late+Empire.htm. Accessed 30 Oct 2011.

Hittites, http://www.jewishvirtuallibrary.org/jsource/judaica/ejud_0002_0009_0_09061.html. Accessed 17 Sept 2011.

The Hittite Empire, http://www.specialtyinterests.net/hittites.html. Accessed 18 Sept 2011.

Plague prayers of the Hittite King Mursili, http://mcroberts-robert.suite101.com/the-plague-prayers-of-mursili-ii-a139219. Accessed 3 Dec 2013.

Taracha, P. (2009). Hittitology in Warsaw: Past, present and future. *Rocznik Orientalistyczny, T.LXII*(Z. 1.), 213–221.

The Hittites, http://www.michaelmaxwolf.de/antike/alter_orient/hethiter.htm in this German article the author talks about the Hittite Empire in general and more importantly about the 1322 BCE Hittite epidemic and states that it was caused specifically by bubonic plague. This is contrary to the current belief that this epidemic was caused by smallpox. This book also supports the hypothesis that the 1322 BCE Hittite epidemic was caused by bubonic plague. Accessed 3 Dec 2013.

The Hittites in history, http://starways.net/lisa/essays/hittites.html. Accessed 3 Dec 2013.

The Hittites of Anatolia, http://www.saudiaramcoworld.com/issue/199405/in.search.of.the.past-the.hittites.of.anatolia.htm. Accessed 3 Dec 2013.

The Zannanza affair, http://realhistoryww.com/world_history/ancient/Misc/Egypt/a_The_Zannanza_Affair.htm Accessed 30 October 2011. This short article discussed the Plague Prayers of the Hittite King Mursili in general terms and did not actually have any of the prayers included, unlike Singer's book which contained all the prayers in their entirety. This article provided a good background read. Accessed 3 Dec 2013.

(4) *Articles About the Sea Peoples*

The Sea Peoples, http://www.sjsu.edu/faculty/watkins/seapeoples.htm. Accessed 3 Dec 2013.

The Sea People and their migration, http://www.american-buddha.com/bible.seapeopleshellpeczynski2.htm. Accessed 1 Jan 2012.

(5) *Articles About the Current Theories as to the Causes of the End of The Bronze Age*

Aegean Bronze Age, http://www.therafoundation.org/articles/history/aegean-bronzeage. This one page article mentioned that famines and pestilence existed at the time of the Minoan collapse, which is just before the time frame of this book. It also put a lot of emphasis on the breakdown of trade routes as a cause for the collapse. Accessed 30 Oct 2011.

Anatolian tree rings and a new chronology for the East Mediterranean Bronze—Iron Ages, http://www.sciencemag.org/294/5551/2532.abstract. Expert evidence from Kuniholm. Accessed 25 Apr 2012.

Anatolian tree rings and the absolute chronology of the Mediterranean, 2220–718 BC, http://www.nature.com/nature/journal/v381/n6585/abs/381780a0.html this article by Kuniholm provided excellent primary evidence of a drought at the beginning of the end of the Bronze Age. He gathered this evidence by examining the tree rings of ancient trees from all over Anatolia. He and his Aegean Dendrochronology Project at the Department of the History of Art and Archaeology at Cornell University are now famous as a reliable resource about dating various events around the Aegean. This article was used in this book. Accessed 3 Dec 2013.

Archaeological evidence and non-evidence for climate change, http://rsta.royalsocietypublishing.org/content/330/1615/645.short. Accessed 3 Dec 2013.

Bronze Age Aegean, http://web.ics.purdue.edu/~rauhtopicn/BAaegeantext.htm this article was from the Purdue University history course on the Bronze Age collapse and the comments are the same as above. Accessed 3 Dec 2013.

Bronze Age collapse, http://www.enotes.com/topic/Bronze_Age_collapse this three page article listed seven causes for the end of the Bronze Age, whereas this book lists sixteen possible causes. It did not mention disease at all and but it was a good general background read which was used in the book. Accessed 30 Oct 2011.

Contribution of drought to the collapse of the Hittite Empire, http://ie499.yeralan.org/index.php?option=com_content&view=article&id=322:hititler&catid=53:projects&Itemid=61. Accessed 18 Sept 2011.

Donald Hopkins, http://www.cartercenter.org/news/experts/donald_hopkins.html. Accessed 3 Dec 2013.

Herodotus 1.94, The drought ca.1200 BC, and the origin of the Etruscans, http://www.jstor.org/pss/4436222. This article also states that the Sea Peoples may have been 'pushed' out of their homelands by drought and resultant famine. Accessed 2 Oct 2011.

Kuniholm, P. I. et al. *Dendrochronological dating in Anatolia: The second millennium BC* (pp. 41–46). The Malcolm and Carolyn Wiener Laboratory for Aegean and Near Eastern Dendrochronology, Cornell University. This article is similar to the one below by Kuniholm and was also used in this book.

Lesson 28: The collapse of the Mycenaean palatial civilization, http://projectsx.dartmouth.edu/history/bronze_age/lessons/les/28.html. This article is part of the prestigious Dartmouth University's History course. It is along sixteen page article which has in the conclusion section five possible causes for the collapse of the Mycenaean Palatial Civilizations, they being economic factors, climate change, internal social upheaval, invasion by the Sea Peoples and finally changes in warfare. There was no mention of disease at all. By contrast this book lists sixteen possible causes for the end of the Bronze Age. Accessed 25 Sept 2011.

Results of climate change circa 1200 BC, http://2012forum.com/forum/viewtopic.php?f=48&t=17630 this article backs up the findings of Kuniholm above by showing that drought occurred at the beginning of the end of the Bronze Age around 1200 BCE. Accessed 3 Dec 2013.

Scientist links killer quakes to Bronze Age's end, http://www.trussel.com/prehist/news46.htm this press release is about Nur, from the Geophysics Department at Stanford University, who said that the end of the Bronze Age was due to storms of killer earthquakes which destroyed the palatial cities and their empires. It does not mention disease at all. Accessed 3 Dec 2013.

The catastrophe—Part 2: What the end of Bronze Age civilization means for modern times, http://www.brusselsjournal.com/node/4106. Accessed 3 Dec 2013.

The collapse of the Bronze Age, http://teachingcompany.fr.yuku.com/topic/2466#.Up-qGigcg04 this seven page article mentions disease and a possible pandemic but then states that this is not supported by the evidence. It also states that any deaths by famine or disease were the result of war rather than famine and disease being the actual cause of the collapse of the Bronze Age. This book refutes all these claims. Accessed 3 Dec 2013.

The crisis or the end of the Bronze Age Part I, http://makinapacalatxilbalba.blogspot.com/2009/01/crisis-or-end-of-bronze-age-part-i-end.html. Accessed 3 Dec 2013.

The influence of climate change on the Late Bronze Age collapse and the Greek Dark Ages, http://academia.edu/1411970/The_Influence_of_Climate_Change_on_the_Late_Bronze_Age_Collapse_and_the_Greek_Dark_Ages this article provides good scientific evidence that the Late Bronze Age was subjected to drought, which may have helped precipitate the Collapse. Accessed 3 Dec 2013.

Thera, Tin, and the Aryan invasion, http://ancienthistroy.about.com/library/bl/uc-slattery_tin.htm. Accessed 3 Dec 2013.

What caused the end of the Bronze Age? http://www.exampleessays.com/viewpaper/1587.html this one page article argues that a shortage of tin, resulting in a shortage of bronze, was the reason for the end of the Bronze Age. It states that tin was scarce and had to come from as far away as Cornwall to supply demand. If these tin trade routes were adversely affected then the supply of tin would dry up thus ending the Bronze Age and people would have to look for an alternative metal—iron. Accessed 3 Dec 2013.

(6) *Articles About Diseases That Could Have Caused the End of The Bronze Age*

Archaeology and the conquest of the land, http://www.testimonymagazine.org/back/Special_Issues/Oct90/art5.htm. Accessed 3 Dec 2013.

Arnott, R. (1996). Healing and medicine in the Aegean Bronze Age. *Journal of the Royal Society of Medicine, 89*, 265–270.

Bubonic plague originated in China, http://news.discovery.com/human/black-death-plague-china.html. Accessed 3 Dec 2013.

Conrad, L. (1984). The biblical tradition for the plague of the Philistines. *Journal of the American Oriental Society, 104*(2), 281–287. This article argues that the Plague of the Philistines was due to bubonic plague.

Cooray, M. P. M. (1965). Epidemics in history. *Ceylon Medical Journal, June–Sept*, 88–96.

Diffusions from Mesopotamia to Egypt, http://www.fsmitha.com/hl/ch03-hyk.html. Accessed 24 Sept 2011.

Hittites used germ warfare, http://www.telegraph.co.uk/news/worldnews/1571927/Hittites-used-germ-warfare-3500–years-ago this newspaper article states that the Hittites used tularemia infected sheep as the first form of germ warfare 3,200 years ago based on the research of Trevisanato. Accessed 24 Sept 2011.

Horwitz, L. K. (2000). The contribution of animal domestication to the spread of zoonoses: A case study from the Southern Levant. *Ibex Journal of Mountain Ecology, 5*(2000), 77–84.

Infectious diseases in history, http://urbanrim.org.uk/diseases.html. Accessed 20 Nov 2011.

Middleton, G. D. (2008). *The collapse of the palatial society in LBA Greece and the postpalatial period*. Ph.D. thesis, Durham University. http://etheses.dur.ac.uk/2900/ this was a very thorough and useful examination of the topic. Plagues were discussed in chapter 3 'Theories of Mycenaean Collapse' but were dismissed for lack of evidence, which this book will supply. Accessed 3 Dec 2013.

On the origin of smallpox: Correlating variola phylogenics with historical smallpox records, http://www.pnas.org/content/104/40/15787.full. Accessed 3 Dec 2013.

Origin of bubonic plague, http://creation.com/the-origin-of-bubonic-plague. Accessed 3 Dec 2013.

Origins of smallpox, http://ruleof6ix.fieldofscience.com/2011/03/on-origins-of-smallpox-where-and-when.html. Accessed 3 Dec 2013.

Ramesses, V. http://famouspharaohs.blogspot.com/2009/02/rameses-v-c-1148-1144-bc.html. Accessed 8 Oct 2011.

Sabbatani, S. (2010). The plague of the Philistines and other pestilences in the Ancient World: exploring relations between the religious-literary tradition, artistic evidences and scientific proofs. *Le Infezioni in Medicina*, *3*, 199–207. In this article Sabbatani argues that the plague of the Philistines was due to bubonic plague.

Shrewsbury, J. F. D. (1949). The plague of the Philistines. *London: Journal of Hygiene*, *47*(3), 244–252. This is an excellent article in which Shrewsbury gives several possible diseases as the cause for the end of the Bronze Age from bubonic plague to dysentery and including hemorrhoids. This article was quoted several times during the book.

Smallpox, http://www.gorydetails.net/demo_sites/SmallpoxSite/smpx_history01.html. Accessed 3 Dec 2013.

Smallpox, http://www.infoplease.com/cig/dangerous-diseases-epidemics/smallpox-1200-years-terror.html. Accessed 10 Oct 2011.

The Philistines 1, http://www.bga.nl/en/articles/filist1.html. Accessed 3 Dec 2013.

The Philistines, http://www.sacred-texts.com/ane/phc/phc05.htm. Accessed 21 Mar 2012.

Trevisanato, S. I. (2004). Did an epidemic of Tularemia in Ancient Egypt affect the course of world history? *Medical Hypotheses*, *63*, 905–910. In this article Trevisanato, an Italian molecular biologist, discusses how a disease entered the main harbor of ancient Egypt called Avaris and killed many people except the Hebrew population. This disease resistance gave Trevisanato the idea that it may have been tularemia.

Trevisanato, S. I. (2007a). The 'Hittite plague', an epidemic of tularemia and the first record of biological warfare. *Medical Hypotheses*, *69*, 1371–1374. In his third and final article Trevisanato argues that the Hittite epidemic of 1322 BCE was also due to tularemia and not smallpox as had been previously thought.

Trevisanato, S. I. (2007b). The biblical plague of the Philistines now has a name, tularemia. *Medical Hypotheses*, *69*, 1144–1146. In the second of Trevisanato's trilogy of articles on the infectious disease tularemia he argues that tularemia was the disease described in the Plague of the Philistine's and not anthrax as had been previously thought.

Tuberculosis, http://www.essortment.com/diseases-illnesses-history-tuberculosis-21066.html. Accessed 3 Oct 2011.

Tularemia, http://www.ncbi.nlm.nih.gov/pubmedhealth/PMH0001859/. Accessed 3 Dec 2013.

Walloe, L. (1999). Was the disruption of the Mycenaean world caused by repeated epidemics of bubonic plague? *Opusula Atheniensia*, *24*, 121–126. This was an excellent article and was used in the book. It focused only on bubonic plague and predated the research of Panagiotakopulu so was unable to provide any hard evidence to support his theory.

Watson-Williams, E. (1962). The end of an epoch. *Greece and Rome*, *9*(2), 109–125. Comments are the same as made above.

Yersinia pestis genome sequencing identifies patterns of global phylogenetic diversity, http://www.nature.com/ng/journal/v42/n12/full/ng.705.html. Accessed 22 Apr 2012.

(C) Special Correspondence

Trevor Bryce—Hittitologist at University of Queensland—t.bryce@uq.edu.au

Liz Rouse—Head Librarian at the Medical History Library at The Royal Australasian College of Physicians—racplib@racp.edu.au

Martin Killeen—Rare Books Librarian, Cadbury Research Library, Special Collections, Academic Services, University of Birmingham and the Barnes Library at the Medical School, University of Birmingham—for information about the W. H. Wynn collection of his notes—M.Killeen@bham.ac.uk

Peter Basham—Historical Collections, Royal College of Physicians—Historical.Services@rcplondon.ac.uk

Edward Bishop at Wellcome Library—e.bishop@wellcome.ac.uk

Index[1]

[1] *Page numbers in bold refers images and tables*

P. Norrie, *A History of Disease in Ancient Times*,
DOI 10.1007/978-3-319-28937-3

The manufacturer's authorised representative in the EU is Springer Nature Customer Service Centre GmbH, Europaplatz 3, 69115 Heidelberg, Germany. If you have any concerns regarding our products, please contact ProductSafety@springernature.com

Printed and bound by CPI Group (UK) Ltd, Croydon, CR0 4YY
30/09/2026
02225514-0013